Contents

Getting the most from this book 4
About this book 5

Content Guidance

Thermodynamics 6
Rate equations 20
Equilibrium constant K_p for homogeneous systems 32
Electrode potentials and electrochemical cells 39
Acids and bases 52

Questions & Answers

Thermodynamics 73
Rate equations 78
Equilibrium constant K_p for homogeneous systems 81
Electrode potentials and electrochemical cells 84
Acids and bases 89

Knowledge check answers 93
Index 94

Getting the most from this book

Exam tips

Advice on key points in the text to help you learn and recall content, avoid pitfalls, and polish your exam technique in order to boost your grade.

Knowledge check

Rapid-fire questions throughout the Content Guidance section to check your understanding.

Knowledge check answers

1 Turn to the back of the book for the Knowledge check answers.

Summaries

- Each core topic is rounded off by a bullet-list summary for quick-check reference of what you need to know.

Exam-style questions

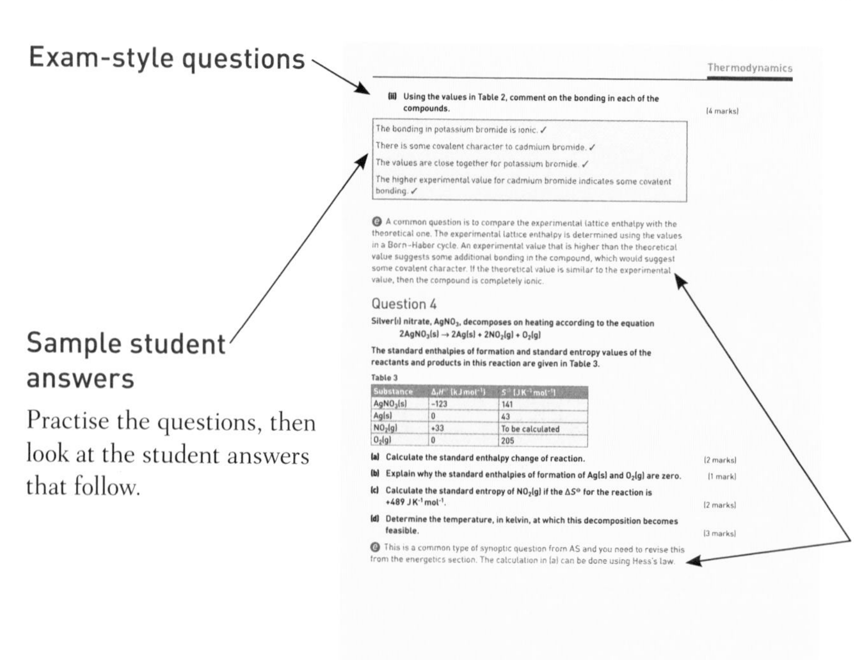

Thermodynamics

(iii) Using the values in Table 2, comment on the bonding in each of the compounds. (4 marks)

The bonding in potassium bromide is ionic. ✓

There is some covalent character to cadmium bromide. ✓

The values are close together for potassium bromide. ✓

The higher experimental value for cadmium bromide indicates some covalent bonding. ✓

ⓔ A common question is to compare the experimental lattice enthalpy with the theoretical one. The experimental lattice enthalpy is determined using the values in a Born–Haber cycle. An experimental value that is higher than the theoretical value suggests some additional bonding in the compound, which would suggest some covalent character. If the theoretical value is similar to the experimental value, then the compound is completely ionic.

Question 4

Silver(I) nitrate, $AgNO_3$, decomposes on heating according to the equation

$$2AgNO_3(s) \rightarrow 2Ag(s) + 2NO_2(g) + O_2(g)$$

The standard enthalpies of formation and standard entropy values of the reactants and products in this reaction are given in Table 3.

Table 3

Substance	$\Delta_f H^\ominus$ (kJ mol^{-1})	$S^\ominus$ (J K^{-1} mol^{-1})
$AgNO_3(s)$	−123	141
$Ag(s)$	0	43
$NO_2(g)$	+33	To be calculated
$O_2(g)$	0	205

(a) Calculate the standard enthalpy change of reaction. (2 marks)

(b) Explain why the standard enthalpies of formation of $Ag(s)$ and $O_2(g)$ are zero. (1 mark)

(c) Calculate the standard entropy of $NO_2(g)$ if the $\Delta S^\ominus$ for the reaction is +489 J K^{-1} mol^{-1}. (2 marks)

(d) Determine the temperature, in kelvin, at which this decomposition becomes feasible. (3 marks)

ⓔ This is a common type of synoptic question from AS and you need to revise this from the energetics section. The calculation in (a) can be done using Hess's law.

Physical chemistry 2 75

Sample student answers

Practise the questions, then look at the student answers that follow.

Commentary on sample student answers

Find out how many marks each answer would be awarded in the exam and then read the comments (preceded by the icon ⓔ) following each student answer showing exactly how and where marks are gained or lost.

About this book

This book will guide you through sections 3.1.8 to 3.1.12 of the AQA A-level Chemistry specification. The sections covered are all physical chemistry. The year 1 physical sections 3.1 to 3.7 are covered in the first student guide of this series.

Paper 1 of A-level covers physical chemistry (3.1.1 to 3.1.12), except 3.1.5 Kinetics and 3.1.9 Rate equations, as well as all inorganic chemistry (3.2.1 to 3.2.6, which can be found in the second and fourth student guides of this series).

Paper 2 of A-lèvel covers organic chemistry (3.3.1 to 3.3.16, found in the second and fourth student guides of this series) as well as 3.1.5 Kinetics and 3.1.9 Rate equations which are covered in the first student guide of this series and this book respectively.

Paper 3 covers all content.

This book has two sections:

- The **Content Guidance** covers the A-level physical chemistry sections 3.1.8 to 3.1.12 and includes tips on how to approach revision and improve exam technique. Do not skim over these tips as they provide important guidance. There are also knowledge check questions throughout this section, with answers at the back of the book. At the end of each section there is a summary of the key points covered. Many topics in the physical chemistry sections covered in the first student guide of this series form the basis of synoptic questions in A-level papers. There are three required practicals related to the topics in this book and notes to highlight these are included.
- The **Questions & Answers** section gives sample examination questions of the types you will find in the exams on each topic, as well as worked answers and comments on the common pitfalls to avoid. This section contains many different examples of questions but you should also refer to past papers, which are available online.

The Content Guidance and the Questions & Answers section are divided into the topics outlined by the AQA A-level specification.

Content Guidance

Thermodynamics

Born–Haber cycles

A Born–Haber cycle is an extension of Hess's law for the formation of an ionic compound. It allows the calculation of lattice enthalpy values.

Lattice enthalpy is most often the enthalpy of lattice dissociation and it is the enthalpy change when 1 mole of an ionic compound is converted into its constituent gaseous ions. The enthalpy of lattice dissociation is an endothermic process. It is represented by $\Delta_L H^\ominus$ or $\Delta_{latt} H^\ominus$.

The enthalpy of lattice formation is an exothermic process and has the same numerical value as the enthalpy of lattice dissociation, but it has a negative sign.

Lattice enthalpy values may be calculated using a Born–Haber cycle.

Born–Haber cycle

The Born–Haber cycle is a technique for applying Hess's law to the standard enthalpy changes that occur when an ionic compound is formed.

The formation of an ionic compound, for example potassium chloride (KCl), may be thought of as occurring in a series of steps, even though the reaction itself may not follow this route.

For KCl, the equation representing the enthalpy change of formation is:

$K(s) + \frac{1}{2}Cl_2(g) \rightarrow KCl(s)$

The **standard enthalpy change of formation** is represented by $\Delta_f H^\ominus$.

The important energy change that we are often trying to determine is the **standard lattice enthalpy** of an ionic compound. This value cannot be determined experimentally, so it must be calculated using the Born–Haber cycle.

For KCl, the equation representing the enthalpy of lattice dissociation is:

$KCl(s) \rightarrow K^+(g) + Cl^-(g)$

The standard lattice enthalpy is represented by $\Delta_L H^\ominus$ or $\Delta_{latt} H^\ominus$.

The Born–Haber cycle linking these enthalpy changes for KCl can be drawn simply as shown in Figure 1.

Exam tip

For all formations, the elements are written in their standard states at 25°C and 1 atm pressure, so K(s) and $Cl_2(g)$. The equation representing the formation must be written for the formation of 1 mole of the compound, in this case KCl(s). Always include state symbols.

The **standard enthalpy change of formation** is the enthalpy change when 1 mole of a compound is formed from its elements in their standard states under standard conditions.

The **standard lattice enthalpy** is the enthalpy change when 1 mole of an ionic compound is converted into gaseous ions.

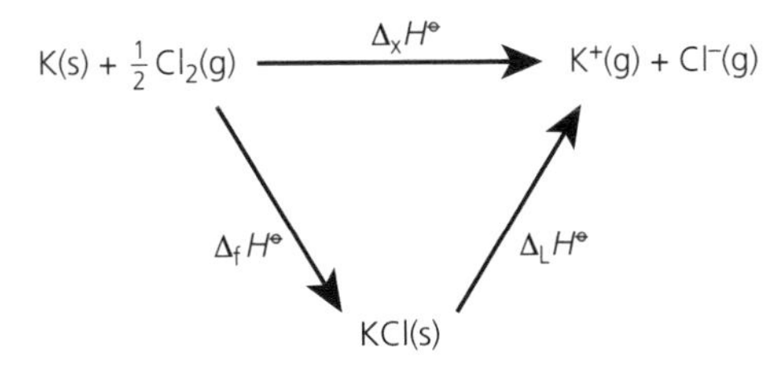

Figure 1 Simple Born–Haber cycle for potassium chloride (KCl)

All enthalpy changes can be determined experimentally apart from the standard lattice enthalpy. $\Delta_x H^\ominus$ is a combination of a few standard enthalpy changes which change $K(s) + \frac{1}{2}Cl_2(g)$ to $K^+(g) + Cl^-(g)$.

Remember, as with any Hess's law diagram, any unknown enthalpy change may be calculated if all the other values are known.

Other enthalpy changes

The change $K(s) + \frac{1}{2}Cl_2(g) \rightarrow K^+(g) + Cl^-(g)$ is composed of the following steps:

Step 1 $K(s) \rightarrow K(g)$

This is the atomisation of potassium. The **standard enthalpy change of atomisation** of potassium is represented by $\Delta_a H^\ominus$ or $\Delta_{at} H^\ominus$.

Step 2 $K(g) \rightarrow K^+(g) + e^-$

This is the first ionisation of potassium. The **first ionisation energy** (enthalpy) is represented by $\Delta_{IE1} H^\ominus$.

Step 3 $\frac{1}{2}Cl_2(g) \rightarrow Cl(g)$

This is the atomisation of chlorine. The same symbol is used for the standard enthalpy change of atomisation as for step 1 for potassium and the definition is the same.

For diatomic elements such as chlorine, the bond dissociation enthalpy may be used.

This enthalpy change is also half of the bond dissociation enthalpy (energy). For diatomic elements the bond dissociation enthalpy (sometimes called the bond enthalpy or bond energy) can be given. If only 1 mole of atoms is required the enthalpy change is half of the bond dissociation enthalpy.

For the change $Cl_2(g) \rightarrow 2Cl(g)$, the enthalpy change is equal to the bond dissociation enthalpy or twice the standard enthalpy change of atomisation.

The **bond dissociation enthalpy** is represented by $\Delta_{BDE} H^\ominus$.

Exam tip

Remember that the bond dissociation enthalpy is twice the standard enthalpy of atomisation for diatomic elements. You may need to use one times the standard enthalpy of atomisation (if 1 mole of atoms is required) or two times the standard enthalpy of atomisation (if 2 moles of atoms are required). If 1 mole of atoms is required you will need to use half of the bond dissociation enthalpy or if 2 moles of atoms are required use one bond dissociation enthalpy.

The **standard enthalpy change of atomisation** is the enthalpy change when 1 mole of gaseous atoms is formed from the element in its standard state under standard conditions.

Knowledge check 1

What is the definition of standard enthalpy change of atomisation?

The **first ionisation energy** is the energy required to remove 1 mole of electrons from 1 mole of gaseous atoms to form 1 mole of gaseous monopositive ions.

The **bond dissociation enthalpy** is the energy required to break 1 mole of a covalent bond under standard conditions.

Step 4 $Cl(g) + e^- \rightarrow Cl^-(g)$

This is the **first electron affinity** of chlorine. It is represented by $\Delta_{EA1}H^\ominus$.

Exam tip

For halides of group 2 elements, two atomisations and two first electrons affinities are required. For example $Cl_2(g)$ is converted to $2Cl(g)$ and $2Cl(g)$ is converted to $2Cl^-(g)$.

The **first electron affinity** is the enthalpy change when 1 mole of electrons is added to 1 mole of gaseous atoms to form 1 mole of gaseous mononegative ions.

A typical Born–Haber cycle diagram

The Born–Haber diagram shown in Figure 2 is that for a typical group 1 halide. This is the Born–Haber cycle for potassium chloride (KCl).

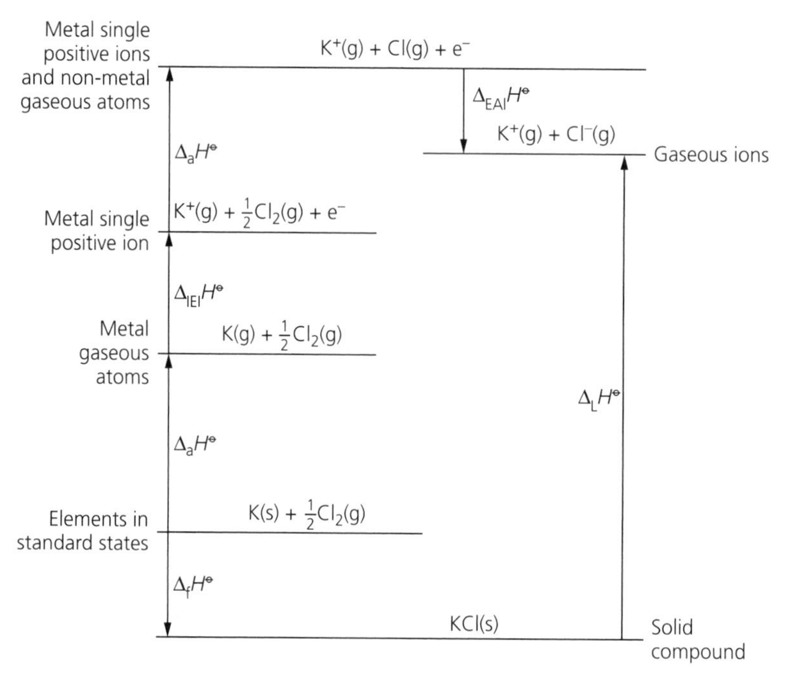

Figure 2 Born–Haber cycle for potassium chloride (KCl)

The values given in a calculation may be:

- enthalpy of formation of potassium chloride ($\Delta_f H^\ominus$) $= -437\,kJ\,mol^{-1}$
- enthalpy of atomisation of potassium ($\Delta_a H^\ominus$) $= +89\,kJ\,mol^{-1}$
- first ionisation energy of potassium ($\Delta_{IE1}H^\ominus$) $= +420\,kJ\,mol^{-1}$
- enthalpy of atomisation of chlorine ($\Delta_a H^\ominus$) $= +121\,kJ\,mol^{-1}$
- first electron affinity of chlorine ($\Delta_{EA1}H^\ominus$) $= -364\,kJ\,mol^{-1}$

$$\Delta_L H^\ominus = -\Delta_f H^\ominus_{(KCl)} + \Delta_a H^\ominus_{(K)} + \Delta_{IE1}H^\ominus_{(K)} + \Delta_a H^\ominus_{(Cl)} + \Delta_{EA1}H^\ominus_{(Cl)}$$

$$= +437 \quad +89 \quad +420 \quad +121 \quad +(-364)$$

$$= +703\,kJ\,mol^{-1}$$

Knowledge check 2

Write an equation with state symbols to represent the enthalpy of lattice dissociation of calcium oxide.

Exam tip

Endothermic processes have upwards arrows and exothermic processes have downwards arrows. There are different forms of this type of diagram, but this is the most common. You will most often be asked to complete the diagram or to use it. Don't forget the electrons or the state symbols. This type of diagram can be applied to any group 1 halide or hydride.

The cycle works from the beginning of the arrow for lattice enthalpy to the end of the arrow. The alternative route gives the same energy changes as predicted by Hess's law. The alternative route must take into account the direction of the arrows. If the direction is reversed, then the negative of the value must be used.

Exam tip

The enthalpy of atomisation of chlorine is sometimes given or the bond dissociation enthalpy can be given. It is vital that you understand that for diatomic elements like the halides, the bond dissociation enthalpy is twice the enthalpy of atomisation. If 2 moles of halide ion are required as shown in the next example, two enthalpies of atomisation of F are required (or one bond dissociation enthalpy) followed by two first electron affinities. The group 2 metal ion requires a first and a second ionisation energy.

This is a standard Born–Haber diagram for all group 1 halides. 1 mole of the group 1 metal, A, reacts with half a mole of the halide, $\frac{1}{2}X_2$, to form 1 mole of the solid halide, AX (s). You may be asked to label the species present at each level in a Born–Haber cycle. Always think about the change that is happening and don't forget to include the electron(s). Note the 1 mole of electrons only appears on two levels for a group 1 halide.

Born–Haber cycle for group 2 halides

For group 2 halides, MX_2, the Born–Haber diagram is slightly extended. Figure 3 shows the Born–Haber cycle for calcium fluoride (CaF_2).

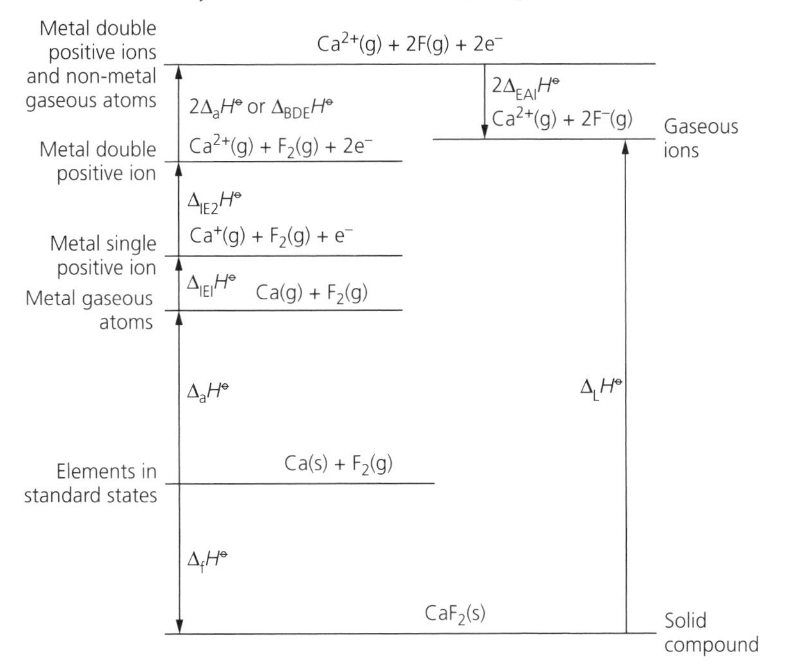

Figure 3 Born–Haber cycle for calcium fluoride (CaF_2)

Exam tip

The main difference to note here is that the first and second ionisation energies of the group 2 metal, A, are needed to form M^{2+}. Two enthalpies of atomisation of the halide, X_2, are needed (or one bond dissociation enthalpy) as well as two electron affinities to form 2 moles of the halide ion, X^-. Don't forget to include the electrons or you will lose all the marks for that level.

Worked example

Calculate the lattice enthalpy of calcium fluoride (CaF_2).

Answer

Lattice enthalpy can be calculated since we know the other values in the Born–Haber cycle:

Enthalpy of formation of calcium fluoride ($\Delta_f H^\ominus$) = −1220 kJ mol^{-1}

Enthalpy of atomisation of calcium ($\Delta_a H^\ominus$) = +178 kJ mol^{-1}

First ionisation energy of calcium ($\Delta_{IE1} H^\ominus$) = +590 kJ mol^{-1}

Second ionisation energy of calcium ($\Delta_{IE2} H^\ominus$) = +1145 kJ mol^{-1}

Bond dissociation enthalpy of fluorine ($\Delta_{BDE} H^\ominus$) = +158 kJ mol^{-1}

First electron affinity of fluorine ($\Delta_{EA1} H^\ominus$) = −348 kJ mol^{-1}

$$\Delta_L H^\ominus = -\Delta_f H^\ominus + \Delta_a H^\ominus + \Delta_{IE1} H^\ominus + \Delta_{IE2} H^\ominus + \Delta_{BDE} H^\ominus + 2\Delta_{EA1} H^\ominus$$

(CaF$_2$) (Ca) (Ca) (Ca) (F) (F)

= +1220 +178 +590 +1145 +158 +2(−348)

= +2595 kJ mol^{-1}

Born–Haber cycle for oxides

When dealing with oxides, the oxygen requires a first and a second electron affinity. The first electron affinity of oxygen is −142 kJ mol^{-1} and the second electron affinity is +844 kJ mol^{-1}. The second electron affinity is endothermic as the second electron is being added to an already negative ion so there is repulsion between the O^- and the e^-.

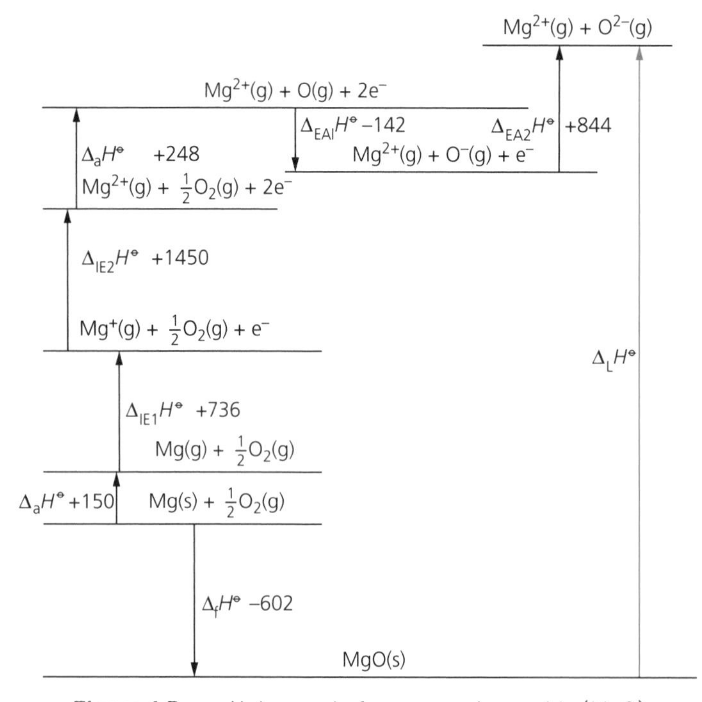

Figure 4 Born–Haber cycle for magnesium oxide (MgO)

For magnesium oxide (MgO):

$$\Delta_L H^\ominus_{(MgO)} = -\Delta_f H^\ominus_{(Mg)} + \Delta_a H^\ominus_{(Mg)} + \Delta_{IE1} H^\ominus_{(Mg)} + \Delta_{IE2} H^\ominus_{(O)} + \Delta_{BDE} H^\ominus_{(O)} + \Delta_{EA1} H^\ominus_{(O)} + \Delta_{EA2} H^\ominus$$

$$= +602 \quad +150 \quad +736 \quad +1450 \quad +248 \quad +(-142) \quad +844$$

$$= +3888\,\text{kJ mol}^{-1}$$

Exam tip

It is expected that the endothermic second electron affinity is shown as endothermic in a Born–Haber cycle so the arrow should go back up and above the level of the $Mg^{2+}(g) + O(g) + 2e^-$.

Patterns in lattice enthalpies

The lattice enthalpy depends on the size of the ions and also the charge on the ions. Table 1 gives the lattice enthalpy values for the group 1 and group 2 halides and oxides.

Table 1 Enthalpy values for group 1 and group 2 halides and oxides

	Anion				
Cation	F^-	Cl^-	Br^-	I^-	O^{2-}
Li^+	+1029	+849	+804	+753	+2808
Na^+	+915	+771	+743	+699	+2481
K^+	+813	+703	+679	+643	+2231
Mg^{2+}	+2883	+2492	+2414	+2314	+3888
Ca^{2+}	+2595	+2197	+2125	+2038	+3523

Looking at the halides of sodium the lattice enthalpy decreases as the size of the halide ion increases (from F^- to I^-). The charge on the ions in these compounds remains the same, but the size of the larger halide ion means the ions cannot pack as closely together, so the attraction between them is not as strong.

A similar pattern is seen by examining the group 1 fluorides. Again the ions are all 1+ and 1–, but the increasing size of the group 1 ion (from Li^+ to K^+) causes the decrease in the lattice enthalpy.

The lattice enthalpies of the group 2 halides shows similar patterns, but they are significantly larger than the lattice enthalpies of the group 1 halides due to the greater attraction between the 2+ ion and the halide ions. Similar patterns are seen for the oxides. Comparing the oxides to the chlorides, the 2– charge on the small oxide ion gives the oxides higher lattice enthalpy values than any of the corresponding halides.

The lattice enthalpy for MgO is very high (and hence it is very stable and has a very high melting point) due to the 2+ and 2– charge on the small ions.

Perfect ionic model

Ionic compounds can show some covalent character. A theoretical value for the lattice enthalpy can be calculated based on the attraction of the ions if the ions are thought of as being perfect spheres with point charges and only electrostatic attraction between the ions.

The experimental value for the lattice enthalpy is determined from the Born–Haber cycle.

If the experimental value and the theoretical value are almost the same the bonding in the compound is ionic.

If the experimental value is greater than the theoretical value this would imply that there is additional bonding in the compound, which would suggest that there is some degree of covalent character.

Exam tip

These points about the perfect ionic model are often asked so it is important to understand the idea that the ions are perfect spheres with point charges with only electrostatic attraction between the ions. The perfect ionic model does not assume any covalent character in the compound.

Worked example

The experimental lattice enthalpy for silver(I) chloride is +905 kJ mol^{-1}, whereas the theoretical value is +770 kJ mol^{-1}. Explain the difference.

Answer

Silver(I) chloride has covalent character, so there is additional bonding in silver(I) chloride and the theoretical value from the perfect ionic model underestimates the lattice enthalpy.

Enthalpy of solution

When an ionic compound dissolves in water the lattice is disrupted and the water molecules form bonds with the ions. The **enthalpy of solution** can be calculated from the lattice enthalpy and the **enthalpy of hydration**.

The **enthalpy of solution** is the enthalpy change when 1 mole of a solute dissolves in water.

The **hydration enthalpy** is the enthalpy change when 1 mole of gaseous ions is converted into 1 mole of aqueous ions.

Hydration enthalpy

Hydration is an exothermic process as energy is released when the bonds form between the polar water molecules and the ions. The $\delta+$ H atoms on water interact with the negative ions and the $\delta-$ O atom interacts with the positive ions. Energy is released.

Hydration enthalpies are per mole of the ion so the hydration enthalpy of magnesium ions is represented by the equation: $Mg^{2+}(g) \rightarrow Mg^{2+}(aq)$. For chloride ions the change would be: $Cl^{-}(g) \rightarrow Cl^{-}(aq)$.

The numerical value for the enthalpy of hydration depends on the size and charge of the ions. Smaller ions with a higher charge have the highest numerical values for the enthalpy of hydration.

For the halides, $F^- = -506\,kJ\,mol^{-1}$; $Cl^- = -364\,kJ\,mol^{-1}$; $Br^- = -351\,kJ\,mol^{-1}$; $I^- = -307\,kJ\,mol^{-1}$. As the size of the ion increases, the enthalpy of hydration decreases in numerical value.

A similar pattern is observed in groups 1 and 2.

Between groups 1 and 2, the group 2 ions have a much higher numerical value for hydration enthalpy than the group 2 ions. For example:

$Na^+ = -407\,kJ\,mol^{-1}$ $\quad Mg^{2+} = -1891\,kJ\,mol^{-1}$.

The Mg^{2+} is smaller and has a higher charge than the Na^+ ion.

Calculating enthalpy of solution

Enthalpy of solution = lattice enthalpy + sum of enthalpies of hydration

For magnesium chloride:

$$\Delta_{sol}H_{(MgCl_2)} = \Delta_L H^\ominus_{(MgCl_2)} + \Delta_{hyd}H^\ominus_{(Mg^{2+})} + 2\Delta_{hyd}H^\ominus_{(Cl^-)}$$

Hydration enthalpies are exothermic and lattice enthalpy is endothermic. It is the balance between lattice enthalpy and the **sum** of all the hydration enthalpies which will determine whether the enthalpy of solution is endothermic or exothermic overall.

These enthalpy changes fit together in the cycle shown in Figure 5.

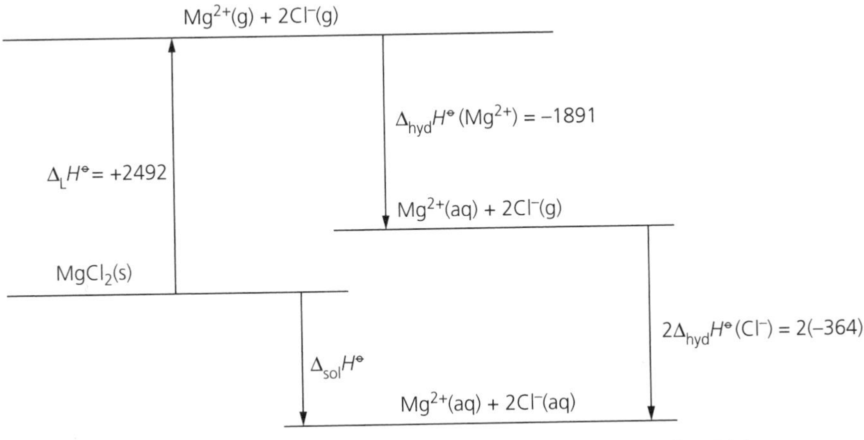

Figure 5 Born–Haber cycle for magnesium chloride ($MgCl_2$)

$$\Delta_{sol}H^\ominus = (+2492) + (-1891) + 2(-364) = -127\,kJ\,mol^{-1}$$

It is clear that 127 kJ of energy are released when 1 mole of $MgCl_2$ dissolves in water to form a solution. All of the hydration enthalpies may be combined into one arrow.

Exam tip

$2\Delta_{hyd}H^\ominus$ are required for magnesium chloride as there are 2 moles of chloride ions being hydrated. As with all enthalpy changes, check carefully the number of moles of the ions being hydrated.

Exam tip

It is important to show the solution ($Mg^{2+}(aq) + 2Cl^-(aq)$) line below the solid ($MgCl_2(s)$) line if the process is exothermic or above it if it is endothermic.

Knowledge check 3

Calculate the enthalpy of solution of sodium iodide given that its lattice enthalpy is $+699\,kJ\,mol^{-1}$ and the hydration enthalpies of the sodium ion and the iodide ion are $-407\,kJ\,mol^{-1}$ and $-307\,kJ\,mol^{-1}$ respectively.

Summary

- Lattice enthalpy is the enthalpy of the breaking of 1 mole of the ionic lattice into gaseous ions and is endothermic (ΔH is positive).
- A Born–Haber cycle allows us to calculate the lattice enthalpy from atomisation enthalpy values, ionisation enthalpy values, bond dissociation enthalpy values, electron affinities and enthalpy of formation values.
- The bond dissociation enthalpy values for halogens is twice the value of the enthalpy of atomisation.
- Second electron affinities are endothermic (ΔH is positive).
- The perfect ionic model allows the calculation of a theoretical value for lattice enthalpy.
- Ionic compounds which have a theoretical value for lattice enthalpy less than the experimental value from a Born–Haber cycle have some covalent character.
- Enthalpy of solution is the total of the lattice enthalpy and the enthalpies of hydration of **all the ions**.

Gibbs free energy change, $\Delta G^\ominus$ and entropy change, $\Delta S^\ominus$

Endothermic reactions are not energetically favourable in terms of the intake of energy, so for them to occur spontaneously or to be feasible, they show an increase in entropy. Entropy is a measure of disorder.

Estimating a change in entropy

Reaction 1: $KI(s) \rightarrow K^+(aq) + I^-(aq)$ $\quad \Delta H^\ominus = +31\,kJ\,mol^{-1}$

Reaction 2: $NH_4Cl(s) \rightarrow NH_3(g) + HCl(g)$ $\quad \Delta H^\ominus = +176\,kJ\,mol^{-1}$

Reaction 3: $H_2O(s) \rightarrow H_2O(l)$ $\quad \Delta H^\ominus = +6\,kJ\,mol^{-1}$

In general, gases are more disordered than liquids and solutions, which are more disordered than solids. This means that the entropy of solids is low. The more ordered the structure of the solid, the lower the entropy. Gases are highly disordered.

The change in entropy in a reaction can be estimated from the state of the substances in a reaction as described previously or from the number of moles of substances should they all be in the same state. Also in general for gases, the higher the M_r of the gas the higher its entropy.

All three reactions above show an increase in entropy.

- Reaction 1 has 1 mole of solid forming 1 mole of each of the ions in the solution.
- Reaction 2 has 1 mole of solid forming 2 moles of gas.
- Reaction 3 has 1 mole of solid forming 1 mole of liquid.

Entropy of changes of state

Changes of state show a change in entropy. This can be shown on a graph of entropy plotted against temperature (Figure 6).

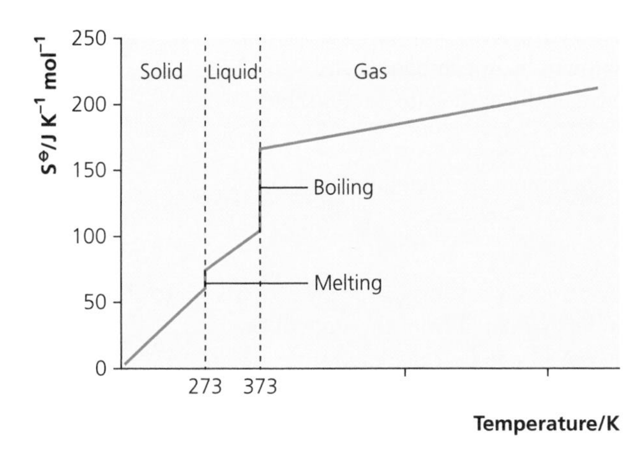

Figure 6 Change in entropy

The four points below are common questions about this type of graph:

- At zero kelvin (0 K) all substance should have an entropy value of zero. This is because at zero kelvin (K) the particles should not be moving and should have perfect order.
- As temperature increases, entropy increases as the substance becomes more disordered as the particles gain energy and movement increases.
- The vertical sections on the graph show the changes in state where there is a large increase in **entropy** as the temperature remains constant.
- The vertical section on boiling is longer than the one for melting because there is a very large increase in entropy when a substance changes into a gas as gases are very disordered.

Entropy is a measure of disorder or randomness in a system.

Entropy values

Entropy is given the symbol S and so standard entropy is $S^{\ominus}$. Change in entropy is represented by $\Delta S^{\ominus}$. An increase in the disorder of a system has a positive $\Delta S^{\ominus}$ value. Entropy is measured in $J\,K^{-1}\,mol^{-1}$.

Some entropy ($S^{\ominus}$) values are given in Table 2 in order of increasing entropy.

Table 2 Entropy of various substances

Substance	State	$S^{\ominus}$ ($J\,K^{-1}\,mol^{-1}$)
C (diamond)	Solid	2.4
Si(s)	Solid	19
Mg(s)	Solid	32.7
CaO(s)	Solid	39.7
$H_2O(s)$	Solid	48
$H_2O(l)$	Liquid	70
$CaCO_3(s)$	Solid	93
NaCl(aq)	Aqueous	116
$C_2H_5OH(l)$	Liquid	161
$H_2O(g)$	Gas	189
$O_2(g)$	Gas	205
$CO_2(g)$	Gas	214

The pattern is not perfect because $CaCO_3(s)$ has a higher entropy value than water. Water has an ordered structure for a liquid owing to hydrogen bonds, and $CaCO_3$ is not a well-ordered ionic solid owing to the size of the carbonate ion and the polarisation of this ion by the calcium ion.

In general, highly ordered structures such as macromolecular, metallic and ionic crystalline structures have low enthalpy values.

Change in entropy, $\Delta S^\ominus$

The change in entropy ($\Delta S^\ominus$) in a reaction can be estimated from the state of the substances in a reaction as described previously. $\Delta S^\ominus$ can be calculated from standard entropy values for a given chemical reaction. A positive value for $\Delta S^\ominus$ indicates that there is an increase in entropy and a negative value indicates that there is a decrease in entropy.

$\Delta S^\ominus$ = sum of $S^\ominus$ for the products – sum of $S^\ominus$ for the reactants

Worked example 1

Calculate the change in entropy ($\Delta S^\ominus$) for the reaction:

$$2SO_2(g) + O_2(g) \rightarrow 2SO_3(g)$$

given the standard entropies, in $J\,K^{-1}mol^{-1}$, of $SO_2(g)$, $O_2(g)$ and $SO_3(g)$ are 248, 205 and 257 respectively.

Answer

$\Delta S^\ominus$ = sum of $S^\ominus$ for the products – sum of $S^\ominus$ for the reactants

$$\Delta S^\ominus = 2 \times 257 - (2 \times 248 + 205) = -187\,J\,K^{-1}\,mol^{-1}$$

Worked example 2

Calculate the entropy value for $H_2(g)$ in the following reaction:

$$CH_4(g) + 2H_2O(g) \rightarrow CO_2(g) + 4H_2(g)$$

if $\Delta S^\ominus = +174\,J\,K^{-1}\,mol^{-1}$ and the standard entropies of $CH_4(g) = 186\,J\,K^{-1}\,mol^{-1}$, $H_2O(g) = 189\,J\,K^{-1}\,mol^{-1}$ and $CO_2(g) = 214\,J\,K^{-1}\,mol^{-1}$.

Answer

$\Delta S^\ominus$ = sum of $S^\ominus$ for the products – sum of $S^\ominus$ for the reactants

Let x = standard entropy of hydrogen in $J\,K^{-1}mol^{-1}$:

$$+174 = (214 + 4x) - (186 + 2 \times 189)$$

$$+174 = -350 + 4x$$

$$4x = 524$$

$$x = 131\,J\,K^{-1}\,mol^{-1}$$

Exam tip

Make sure that the + or – sign is always used when giving changes in entropy, for example $\Delta S^\ominus = +22.5\,J\,K^{-1}\,mol^{-1}$ or $\Delta S^\ominus = -17.4\,J\,K^{-1}\,mol^{-1}$. It is also very important to remember that the units of $S^\ominus$ and $\Delta S^\ominus$ are $J\,K^{-1}\,mol^{-1}$.

Knowledge check 4

What does $\Delta S^\ominus$ mean and what are its units?

Exam tip

For all ΔS calculations the 'per mole' in the equation must be taken into account. If 2 moles of a reactant are used, then 2 moles of the entropy value for this substance are used.

Exam tip

This is a slightly different type of calculation, but uses the same expression to calculate an entropy value from a change in entropy.

Gibbs free energy, $\Delta G^\ominus$

Gibbs free energy is represented by $\Delta G^\ominus$ and is calculated using the expression $\Delta G^\ominus = \Delta H^\ominus - T\Delta S^\ominus$, where $\Delta H^\ominus$ is the standard enthalpy change for the reaction measured in $kJ\,mol^{-1}$; T is the temperature measured in kelvin (K) and $\Delta S^\ominus$ is the standard enthalpy change measured in $kJ\,K^{-1}\,mol^{-1}$. ΔG has units of $kJ\,mol^{-1}$.

In order for a reaction to be feasible $\Delta G^\ominus$ must be zero or negative. You can calculate whether a reaction is feasible at a certain temperature by calculating $\Delta G^\ominus$. You can also calculate the temperature at which a reaction becomes feasible using $\Delta G^\ominus$ and $\Delta H^\ominus$, assuming that $\Delta G^\ominus = 0$ for a feasible reaction.

Worked example

Show that the thermal decomposition of silver(I) nitrate is not feasible at 600 K and determine the temperature at which the reaction becomes feasible.

$$2AgNO_3(s) \rightarrow 2Ag(s) + 2NO_2(g) + O_2(g) \qquad \Delta H^\ominus = +312\,kJ\,mol^{-1}$$

given the following standard entropy values:

Substance	$S^\ominus$ ($J\,K^{-1}\,mol^{-1}$)
$AgNO_3(s)$	141
$Ag(s)$	43
$NO_2(g)$	240
$O_2(g)$	205

Answer

First $\Delta S^\ominus$ is calculated as before using:

$\Delta S^\ominus$ = sum of $S^\ominus$ for the products – sum of $S^\ominus$ for the reactants

$$\Delta S^\ominus = (2 \times 43 + 2 \times 240 + 205) - (2 \times 141) = +\,489\,J\,K^{-1}\,mol^{-1}$$

Then ΔG is calculated using $\Delta G^\ominus = \Delta H^\ominus - T\Delta S^\ominus$ with the values below:

$\Delta H^\ominus = +312\,kJ\,mol^{-1}$, $T = 600\,K$ and $\Delta S^\ominus = +0.489\,kJ\,K^{-1}\,mol^{-1}$

$$\Delta G^\ominus = +312 - 600(+0.489) = +18.6\,kJ\,mol^{-1}$$

$\Delta G^\ominus$ is positive at 600 K, so the reaction is not feasible at this temperature.

At what temperature does it become feasible?

To find the temperature at which the reaction becomes feasible, we need to work out the temperature at which $\Delta G^\ominus = 0$.

When $\Delta G^\ominus = 0$, $T = \Delta H^\ominus/\Delta S^\ominus$, where $\Delta S^\ominus$ should be in units of $kJ\,K^{-1}\,mol^{-1}$.

$$T = \frac{312}{0.489} = 638\,K$$

Exam tip

The most important thing to remember when calculating a value for $\Delta G^\ominus$ is that the $\Delta S^\ominus$ value quoted in $J\,K^{-1}\,mol^{-1}$ must be divided by 1000 to convert to $kJ\,K^{-1}\,mol^{-1}$ before being used in the expression. This is the most common error in these questions.

Knowledge check 5

What are the units of ΔG?

Exam tip

To determine the temperature for a reaction to become feasible, divide ΔH by ΔS, but remember to make sure ΔS is in $kJ\,K^{-1}\,mol^{-1}$.

You should also revise Hess's law calculations from the Energetics section of AS. These often form part of an entropy question at A-level because you may have to use standard enthalpies of formation to determine the enthalpy change using Hess's law.

Physical changes related to $\Delta G^\ominus$

During a physical change of state $\Delta G^\ominus = 0$ as the system is at equilibrium. This means that $\Delta H^\ominus = T\Delta S^\ominus$. This can be used to work out the enthalpy change for this particular change in state or the melting point or boiling point may be determined.

Worked example

The standard entropy values of bromine as a liquid and a gas are 152 and 245 $J K^{-1} mol^{-1}$. The boiling point of bromine is 332 K. Determine the enthalpy change of vaporisation of bromine, $Br_2(l) \rightarrow Br_2(g)$. Give your answer to 3 significant figures.

Answer

$\Delta S^\ominus$ = sum of $S^\ominus$ for the products – sum of $S^\ominus$ for the reactants

$\Delta S^\ominus = 245 - 152 = +93\,J\,K^{-1}\,mol^{-1}$

As this is a change in state, $\Delta G^\ominus = 0$. So $\Delta H^\ominus = T\Delta S^\ominus$. So:

$\Delta H^\ominus = 332 \times 0.093 = +30.9\,kJ\,mol^{-1}$

As expected, the enthalpy of vaporisation is endothermic.

Exam tip

The enthalpy change of vaporisation of bromine could have been given in this question and the boiling point calculated. Try this calculation to ensure you get 'around' 332 K. Rounding may affect the answer. At temperatures below the boiling point the value for $\Delta G^\ominus$ will be positive.

Knowledge check 6

Calculate the boiling point of ethanol if the change $C_2H_5OH(l) \rightarrow C_2H_5OH(g)$ $\Delta S^\ominus = +110\,J\,K^{-1}\,mol^{-1}$ and $\Delta H^\ominus = +38.6\,kJ\,mol^{-1}$. Give your answer to 3 significant figures.

Factors affecting feasibility of a reaction

The feasibility of a reaction depends on whether a reaction is endothermic or exothermic and also on whether there is an increase in entropy or a decrease in entropy. Table 3 shows how these factors affect $\Delta G^\ominus$ and the feasibility of a reaction.

Table 3 Feasibility of a reaction

$\Delta H^\ominus$	$\Delta S^\ominus$	$\Delta G^\ominus$	Feasibility
Negative	Negative	May be positive or negative	Feasible below certain temperatures
Negative	Positive	Always negative	Feasible at any temperature
Positive	Negative	Always positive	Not feasible at any temperature
Positive	Positive	May be positive or negative	Feasible above certain temperatures

- An exothermic reaction with an increase in entropy is feasible at any temperature.
- An endothermic reaction with a decrease in entropy is *not* feasible at any temperature.

Exam tip

The most often used reactions are the ones which are endothermic and which have an increase in entropy as they allow for the calculation of a temperature at which the endothermic reaction becomes feasible. This can be applied to any reaction including physical changes like dissolving.

Graphs of $\Delta G^\ominus$ against T

$\Delta G^\ominus$ varies with temperature. If a graph of $\Delta G^\ominus$ in $kJ\,mol^{-1}$ is plotted against T in K, the graph will be a straight line in the form $y = mx + c$, where the gradient (m) is $-\Delta S^\ominus$ and the intercept (c) with the $\Delta G^\ominus$ axis is $\Delta H^\ominus$. $\Delta S^\ominus = -$ gradient, but the value will be in $kJ\,K^{-1}\,mol^{-1}$, so would be multiplied by 1000 to convert to $J\,K^{-1}\,mol^{-1}$.

There are two possible graphs based on the gradient being positive or negative (Figure 7).

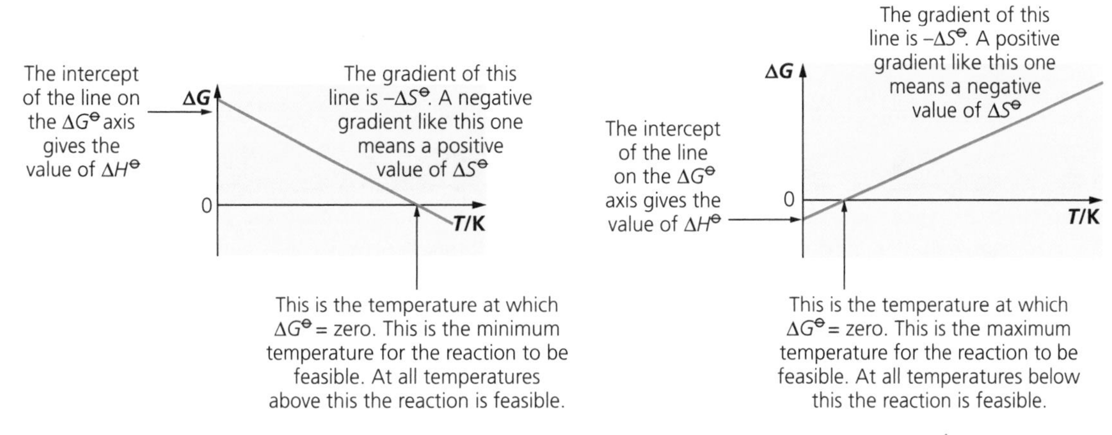

Figure 7 Variation of $\Delta G^\ominus$ with temperature

You would be expected to analyse a graph and calculate a gradient ($-\Delta S^\ominus$) and intercept ($\Delta H^\ominus$) from the graph. A suitable range would be allowed. Figure 8 shows a plot of $\Delta G^\ominus$ against T.

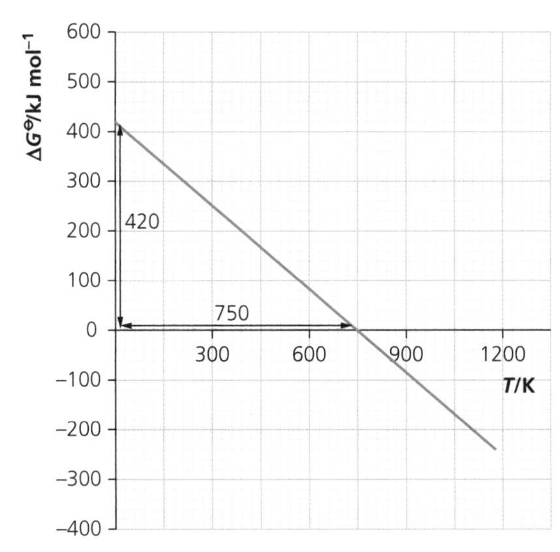

Figure 8 Graph of $\Delta G^\ominus$ against temperature

Exam tip

From the graph, the line intercepts with the $\Delta G^{\ominus}$ axis at +420 kJ mol^{-1}, so this is the value of $\Delta H^{\ominus}$ for this reaction. The gradient is determined by dividing rise over run, which is 420/750 = −0.56 kJ K^{-1} mol^{-1}. The gradient is negative as the line slopes down. $\Delta S^{\ominus}$ = +560 J K^{-1} mol^{-1}. The gradient is $-\Delta S^{\ominus}$ and has to be multiplied by 1000 to convert the units from kJ K^{-1} mol^{-1} to J K^{-1} mol^{-1}. The temperature at which the reaction becomes feasible is the intercept on the T axis, as this is the temperature at which $\Delta G^{\ominus} = 0$. Above this temperature $\Delta G^{\ominus}$ will be negative, so from this temperature upwards the reaction is feasible. This could be checked by calculation using $\Delta G^{\ominus} = \Delta H^{\ominus} - T\Delta S^{\ominus}$.

Summary

- Standard entropy ($S^{\ominus}$) is measured in J K^{-1} mol^{-1}. Entropy is a measure of disorder.
- $\Delta S^{\ominus}$ = sum of the standard entropy values of the products − sum of the standard entropy values of the reactants.
- $\Delta G^{\ominus}$ is Gibbs free energy and is measured in kJ mol^{-1}.
- $\Delta G^{\ominus} = \Delta H^{\ominus} - T\Delta S^{\ominus}$, where T is the temperature in kelvin, $\Delta S^{\ominus}$ is the standard entropy change and $\Delta H^{\ominus}$ is the enthalpy change of the reaction.
- $\Delta S^{\ominus}$ is in J K^{-1} mol^{-1} and it must be divided by 1000 to convert to kJ K^{-1} mol^{-1} for use in the expression $\Delta G^{\ominus} = \Delta H^{\ominus} - T\Delta S^{\ominus}$.
- For a reaction to be feasible $\Delta G^{\ominus}$ must be less than or equal to zero.
- For a change in state $\Delta G^{\ominus} = 0$.
- For a graph of $\Delta G^{\ominus}$ against T, the gradient is $-\Delta S^{\ominus}$ (in kJ K^{-1} mol^{-1}) and the intercept with the $\Delta G^{\ominus}$ axis is equal to $\Delta H^{\ominus}$.

Rate equations

- A rate equation expresses the link between the rate of a chemical reaction and the concentration of the reactants.
- The rate of a reaction can be measured based on how fast the concentration of a reactant is decreasing or how fast the concentration of a product is increasing.
- The units of rate are concentration per unit time, for example, mol dm^{-3} s^{-1} (mol per dm^3 per second).
- Square brackets [ ] as for K_c at AS indicate concentration of the substance inside the brackets. This is always measured in mol dm^{-3} (mol per dm^3). For example:

  $[H^+]$ = concentration of hydrogen ions measured in mol dm^{-3}

  $[I_2]$ = concentration of iodine measured in mol dm^{-3}

The rate equation

The rate equation is an expression showing how the rate of reaction is linked to the concentration of the reactants. Rate is equal to the rate constant (k) multiplied by the concentration of each reactant raised to certain whole number powers (called orders).

The rate equation for the general reaction P + Q → R + S is:

$$\text{rate} = k[P]^x[Q]^y$$

where x is the **order of reaction** with respect to reactant P, y is the order of reaction with respect to reactant Q and k is the **rate constant**.

The **overall order of the reaction** is the sum of all the orders in the rate equation. For this reaction the overall order of reaction would be $x + y$.

The **order of reaction** with respect to a particular reactant is the power to which the concentration of this reactant is raised in the rate equation.

The **rate constant** is the proportionality constant which links the rate of reaction to the concentrations in the rate equation.

The **overall order of a reaction** is the sum of the powers to which the concentration terms are raised in the rate equation.

Units of the rate constant

Rate has units of $mol\,dm^{-3}\,s^{-1}$.

Concentration has units of $mol\,dm^{-3}$.

The units of the rate constant, k, depend on overall order of reaction and the units can be calculated as shown:

For a general rate equation:

rate = k[P][Q]

the overall order is 2 (1 + 1)

rate = $k \times (\text{concentration})^2$

Putting in the units:

$mol\,dm^{-3}\,s^{-1} = k(mol\,dm^{-3})^2$

Rearranging to find k:

$$k = \frac{mol\,dm^{-3}\,s^{-1}}{(mol\,dm^{-3})^2} = \frac{mol\,dm^{-3}\,s^{-1}}{mol^2\,dm^{-6}}$$

Treat each term separately: $mol/mol^2 = mol^{-1}$ and $dm^{-3}/dm^{-6} = dm^{-3-(-6)} = dm^3$.

Units of the rate constant $k = mol^{-1}\,dm^3\,s^{-1}$

Table 4 shows the units of the rate constant for some overall orders.

Table 4 Units of the rate constant

Overall order of reaction	Units of rate constant, k
1	s^{-1}
2	$mol^{-1}\,dm^3\,s^{-1}$
3	$mol^{-2}\,dm^6\,s^{-1}$
4	$mol^{-3}\,dm^9\,s^{-1}$

Exam tip

As the order increases by 1 an extra $mol^{-1}\,dm^3$ is multiplied into the units.

Knowledge check 7

What are the units of the rate constant for the reaction
$2A + B \rightarrow C + 3D$
with the rate equation rate = $k[A][B]^2$?

Determining and using orders of reaction

The order of reaction with respect to the reactants may be determined from experimental data at a constant temperature. Often these data are tabulated.

There may be values missing in the table, which have to be calculated once the orders of reaction have been determined.

Exam tip

The constant temperature is important as changes in temperature affect the rate of the reaction as well.

Worked example 1

For the reaction between two compounds A and B, the following information was obtained from rate experiments at a constant temperature.

Experiment	Initial concentration of A ($mol\,dm^{-3}$)	Initial concentration of B ($mol\,dm^{-3}$)	Initial rate of reaction ($mol\,dm^{-3}\,s^{-1}$)
1	1.20×10^{-4}	2.40×10^{-4}	1.60×10^{-5}
2	1.80×10^{-4}	3.60×10^{-4}	3.60×10^{-5}
3	1.80×10^{-4}	5.40×10^{-4}	8.10×10^{-5}

a Determine the orders of reaction with respect to A and B.
b Write a rate equation for this reaction.
c Deduce a value of the rate constant, k, at this temperature and state its units. Give your answer to three significant figures.

Answer (a)

The key to determining rate is often looking for two experiments in which the concentration of one of the reactants does not change. In this example the concentration of A does not change between experiments 2 and 3. This would be good practice as it allows the effect of the change in concentration of B on the rate to be examined. We then use the factors by which the rate and the concentration have change to determine the order with respect to B.

The rate increased by a factor of $\frac{8.10 \times 10^{-5}}{3.60 \times 10^{-5}} = 2.25$ (factor r — r meaning rate)

The concentration of B increased by a factor of $\frac{5.40 \times 10^{-4}}{3.60 \times 10^{-4}} = 1.5$ (factor b — b meaning concentration of B)

The power to which factor b is raised to equal factor r is the order with respect to B (we are using x to represent the order with respect to B).

$1.5^x = 2.25$ so x must be 2. The order with respect to B is 2.

Now we know the order with respect to one reactant we can determine the order with respect to the other.

The order with respect to A can only be determined using experiments 1 and 2 as this is where the concentration of A is changing. However, the concentration of B is also changing.

The rate increased by a factor of $\frac{3.60 \times 10^{-5}}{1.60 \times 10^{-5}} = 2.25$ (factor r — r meaning rate)

The concentration of A increased by a factor of $\frac{1.80 \times 10^{-4}}{1.20 \times 10^{-4}} = 1.5$ (factor a — a meaning concentration of A)

→

Exam tip

Remember that the orders should be 0, 1 or 2. Try the numbers in your calculator to make sure you can work out that the order is 2.

The concentration of B increased by a factor of $\frac{3.60 \times 10^{-4}}{2.40 \times 10^{-4}} = 1.5$ (factor b — b meaning concentration of B)

As both are changing, the order with respect to B (we are using y this time as the order) is calculated as shown below:

$$(\text{factor } a)^y(\text{factor } b)^2 = \text{factor } r$$

We know the order with respect to B is 2. So

$$(1.5)^y(1.5)^2 = 2.25$$

$$(1.5)^y(2.25) = 2.25$$

$$(1.5)^y = \frac{2.25}{2.25} = 1$$

So $y = 0$.

Answer (b)

The rate equation for this reaction is: rate = $k[B]^2$

As A has zero order it is not in the rate equation for the reaction. It could be included as in rate = $k[A]^0[B]^2$, but it is not necessary as any value to the power of 0 is equal to 1. Zero order reactants can be left out of the rate equation.

Answer (c)

You may use any of the experiments in the table, but sometimes you will be directed to use the results of a particular experiment.

Using experiment 1:

$$\text{rate} = k[B]^2$$

$$1.60 \times 10^{-5} = k(2.40 \times 10^{-4})^2$$

$$k = \frac{1.60 \times 10^{-5}}{(2.40 \times 10^{-4})^2} = \frac{1.60 \times 10^{-5}}{5.76 \times 10^{-8}} = 278\,\text{mol}^{-1}\,\text{dm}^3\,\text{s}^{-1}$$

Exam tip

Any number raised to the power of 0 is equal to 1. Try this on your calculator to make sure you understand.

Worked example 2

Sometimes a value is missing from the table that has to be completed.

In the reaction between X and Y, the following rate measurements were made. The rate equation for the reaction is rate = $k[X]^2[Y]$.

Experiment	Initial concentration of X ($mol\,dm^{-3}$)	Initial concentration of Y ($mol\,dm^{-3}$)	Initial rate of reaction ($mol\,dm^{-3}\,s^{-1}$)
1	2.50×10^{-2}	3.00×10^{-3}	1.45×10^{-3}
2	2.50×10^{-2}	to be calculated	4.36×10^{-3}
3	5.00×10^{-2}	4.50×10^{-3}	to be calculated

The rate equation states that the rate of the reaction is proportional to the concentration of X squared and the concentration of Y. This allows for the calculation of the unknown values in the table.

a Calculate the initial concentration of Y when the rate is $4.36 \times 10^{-3}\,mol\,dm^{-3}\,s^{-1}$.
b Calculate the rate of reaction when the concentration of Y is $4.50 \times 10^{-3}\,mol\,dm^{-3}$.

Answer (a)

Between experiment 1 and 2 the concentration of X stays the same so it has no effect on the rate. The rate changes by a factor of $\frac{4.36 \times 10^{-3}}{1.45 \times 10^{-3}} = 3$ (factor r).
As the order with respect to Y is 1, Y also increases by a factor of 3.

The concentration of Y in experiment 2 is $3.00 \times 10^{-3} \times 3 = 9.00 \times 10^{-3}\,mol\,dm^{-3}$.

Answer (b)

From experiment 1 to experiment 3:
X increases by a factor of $\frac{5.00 \times 10^{-2}}{2.50 \times 10^{-2}} = 2$ (factor x).
Y increases by a factor of $\frac{4.50 \times 10^{-3}}{3.00 \times 10^{-3}} = 1.5$ (factor y).

rate = $k[X]^2[Y]$

factor r = (factor x)^2(factor y)

factor $r = (2)^2(1.5) = 6$

The rate will increase by a factor of 6.

rate = $1.45 \times 10^{-3} \times 6 = 8.70 \times 10^{-3}\,mol\,dm^{-3}\,s^{-1}$

Exam tip

Alternatively the rate constant k could be calculated using experiment 1 ($k = 773\,mol^{-2}\,dm^6\,s^{-1}$). This could be used to calculate the rate for experiment 3 using the rate equation: rate $= 773 \times (5.00 \times 10^{-2})^2 \times (4.50 \times 10^{-3}) = 8.70 \times 10^{-3}\,mol\,dm^{-3}\,s^{-1}$

Factors affecting rate

Rate is directly dependent on concentration of reactants as shown in the rate equation.

rate $\propto$ concentration (rate is proportional to concentration)

Rate is also dependent on temperature and activation energy, but these factors are not seen in the rate equation. The activation energy is affected by the presence of a catalyst. The rate constant, k, is dependent on temperature and activation energy.

Figure 9 shows the effect of temperature on the rate constant.

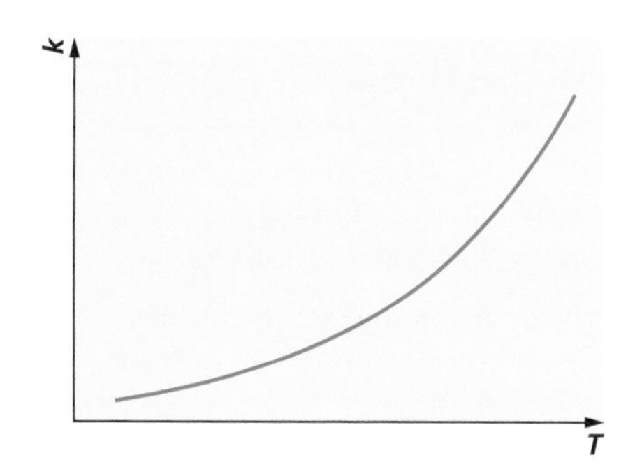

Figure 9 Effect of temperature on the rate constant

Methods of determining rate of reaction

The rate of a chemical reaction may be an **initial rate monitoring** of reaction or it may be **continuous rate monitoring** for measurements taken during the reaction.

- The initial rate is often referred to as the rate at time zero seconds (or $t = 0$ s) when the reaction has just started. At this point the rate is dependent on the concentrations of the reactants added at $t = 0$ s.
- Continuous rate measurements are taken as the reaction proceeds. The rate of reaction decreases as the reaction proceeds, as the reactants are used up.
- Methods of measuring initial rate values relate these to the concentration of the reactant at $t = 0$ s.
- Progressive rate methods can measure the rate at certain points during the reaction and relate this directly to the concentration of reactants at these times.
- This may seem more complex, but it simply relies on being able to measure the concentration of a reactant directly during the reaction.

Rate measurements

Rate measurements are based on a measurable concentration of a reactant or product. The methods depend on the chosen reactant or product. All are measured against time, as their change in concentration against time gives a measure of rate of reaction. The following methods can be used to measure the concentration of a substance against time.

- If the substance is coloured, a colorimeter can be used to measure its concentration. Often a calibration curve is used. This is a previously drawn curve relating absorbance value on the colorimeter to the concentration of known solutions of the coloured substance. It shows a directly proportional relationship. Any absorbance values can be directly converted to a concentration value using the calibration curve.
- If a gas is released, a gas syringe can be used to measure gas volume (which is the same as gas concentration), or the change in mass can be recorded.
- If a substance can be titrated, a sample of the reaction can be taken and quenched and titrated to determine concentration. Quenching may be chemical (removing another reactant) or by cooling the reaction rapidly to stop the reaction at the time when the sample is taken.

Exam tip

When colorimetry is being used to measure the rate of a reaction, a calibration curve using known concentrations of the coloured substance should always be included in the answer.

Exam tip

Substances that can be titrated include acid (using a standard solution of alkali) and alkali (using a standard solution of acid).

- If H^+ or OH^- ions are a reactant or a product, the change in pH can be measured using a pH meter (this can also be titrated with a standard solution of alkali for H^+ ions or a standard solution of acid for OH^-).

Initial rate monitoring

Choose a property that is measurable (this could be gas volume of a product, colour of a product, pH values for a product or a property that can be determined by titration).

For one reactant, choose a range of concentrations of this reactant and set up a series of experiments with a range of concentrations of this reactant.

Plot graphs of measurable quantity against time. This could be gas volume against time, absorbance against time, pH against time or concentration against time determined by titration. Determine the initial gradient to determine initial rate at $t = 0$ s by drawing a tangent at $t = 0$ s and determine its gradient. This is the initial rate of reaction.

Exam tip

You will be measuring the rate of reaction relative to this reactant, as this is the only variable you are changing. So any change in the rate of reaction is caused by the change in concentration of this reactant.

Initial rate of a reaction from a gas volume–time graph

A **gas syringe** is a ground glass syringe which is attached to a sealed reaction vessel and measures the volume of gas produced (Figure 10).

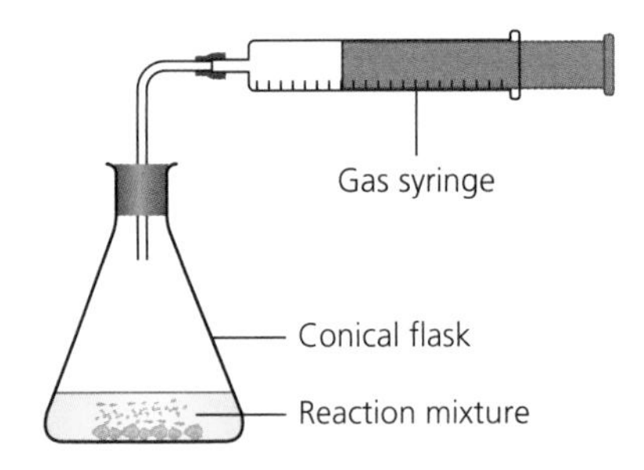

Figure 10 A gas syringe

The volume of gas produced is measured against time and a graph of gas volume against time is plotted (Figure 11).

The initial gradient gives a measure of the initial rate of reaction.

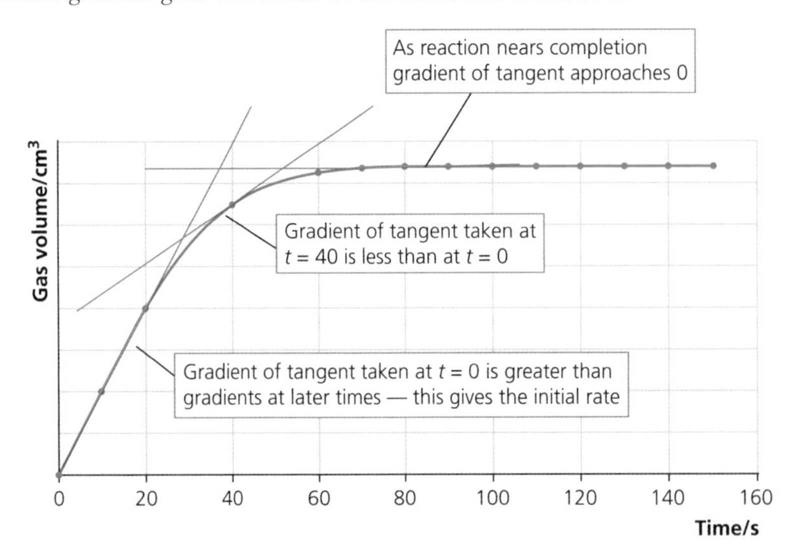

Figure 11 Example of a gas volume–time graph

A reaction in which a gas is produced may also be monitored by measuring the mass over a period of time. The graph decreases and the initial tangent gives a measure of rate of reaction (Figure 12).

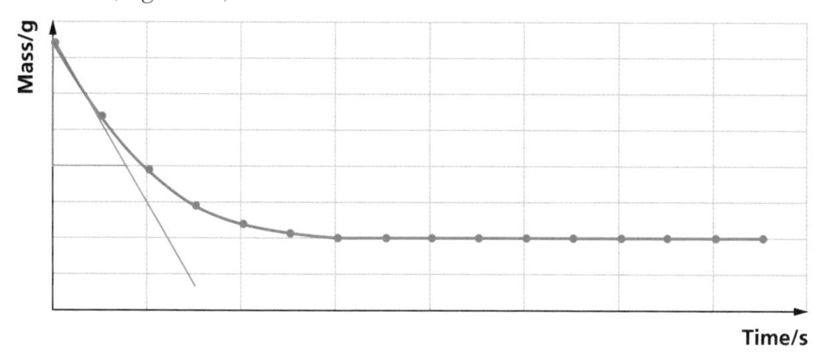

Figure 12 Example of a graph showing change in mass against time

Repeat the experiment for the other concentrations of this reactant. This will give you initial rate values for different concentrations of the reactant you have chosen when you take tangents at $t = 0\,s$. Then plot a graph of initial rate against concentration of this reactant. The shape of the graph of rate against concentration gives the order with respect to this reactant.

Figure 13 shows the graphs expected for order 0, 1 and 2 for the reactant investigated.

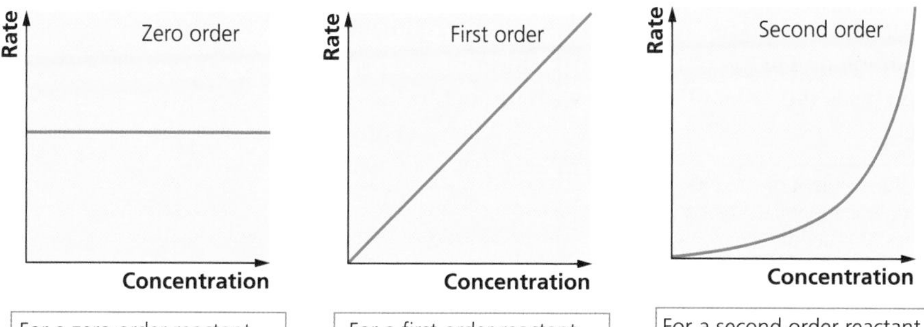

Figure 13 Graphs of rate against concentration

Exam tip

You should be able to recognise and sketch rate versus concentration graphs for these orders of reaction.

Continuous rate monitoring

For this method work out whether the concentration of a reactant can be determined directly. For example:

- Hydrogen ions (H^+) are determined directly from pH measurements. Make sure no other ions would interfere, such as OH^- ions or a different acidic or alkaline product.
- Coloured reactant is determined from colorimeter readings using a calibration curve.
- A specific reactant may be titrated.

Allow the reaction to progress and take readings (colorimeter/pH) or take samples at various times (samples should be quenched to stop the reaction and titrated — quenching can be carried out by rapid cooling or adding large quantities of cold water or chemical quenching). Plot a graph of concentration against time for this reactant. The shape of graph gives the order with respect to this reactant (Figure 14).

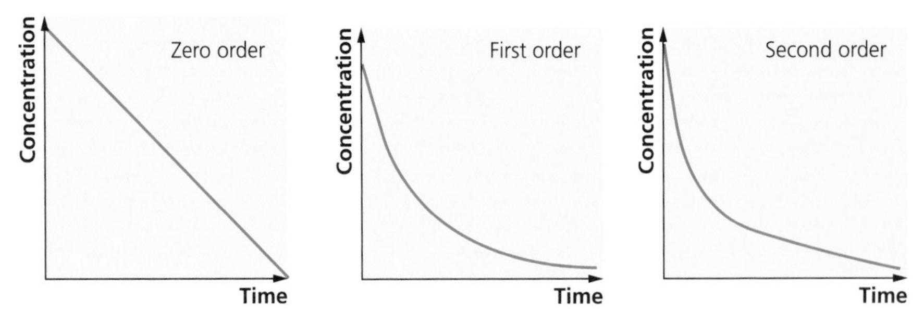

Figure 14 Graphs of concentration against time

The gradient at various points may be taken to determine the rate of reaction. The slope of these graphs (Figure 14) gives the rate (change in concentration against time).

Using a graph of concentration against time

For a coloured reactant we can use continuous rate monitoring. This means one experiment is carried out and the colorimeter readings are converted to concentration using the calibration curve. A graph of concentration against time is drawn and often the shape of this curve can give the order with respect to the coloured reactant (Figure 15). A gradient of a tangent at any concentration on the graph is a measure of rate at that concentration. The rate can then be related to the concentration and the order determined as shown before, or a graph of rate against concentration can be drawn. The shape of this graph gives the order of reaction with respect to the changing reactant concentration.

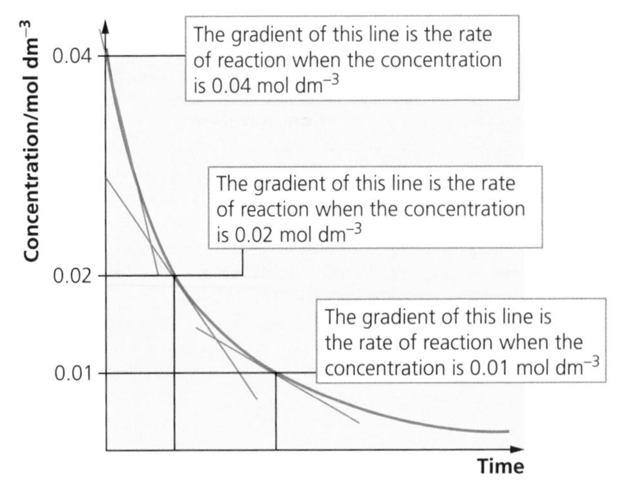

Figure 15 Using a graph of concentration against time

Rate against concentration graphs may be plotted for a clearer order. The values determined from the gradient at different concentrations are used. The graphs of rate against concentration are as shown in initial rate methods and this allows determination of order.

Arrhenius equation

The Arrhenius equation links the rate constant (k) with the activation energy (E_a) and the temperature (T) measured in kelvin. A is the Arrhenius constant and e is a mathematical constant (2.71828).

$$k = Ae^{-E_a/RT}$$

R is the gas constant (8.31 $J\,K^{-1}\,mol^{-1}$) and the activation energy is measured in $J\,mol^{-1}$ in this expression. The Arrhenius constant has the same units as the rate constant.

Worked example

A reaction has an activation energy of 105 $kJ\,mol^{-1}$ at 300 K. The Arrhenius constant $A = 1.27 \times 10^{10}\,s^{-1}$, $R = 8.31\,J\,K^{-1}\,mol^{-1}$, e = 2.71828. Calculate the value of the rate constant k to 3 significant figures.

Answer

$$\frac{E_a}{RT} = \frac{105000}{8.31 \times 300} = 42.12$$

$$k = Ae^{-E_a/RT}$$

$$k = 1.27 \times 10^{10} \times e^{-42.12}$$

$$k = 6.48 \times 10^{-9}\,s^{-1}$$

The units of the rate constant are the same as the units of the Arrhenius constant.

Exam tip

Natural log is the log to the base of e. This means that the log is the power to which e has to be raised to be equal to the number. So ln e = 1, as e would be raised to the power of 1 to be equal to e. e^x is the reverse process of ln, so taking a natural log of e^x is x.

Graphical analysis

The Arrhenius equation can be analysed graphically by plotting a graph of $\ln k$ against $1/T$.

$$\ln k = -\frac{E_a}{RT} + \ln A$$

A graph of $\ln k$ against $1/T$ yields a straight line in the form $y = mx + c$, where the intercept on the $\ln k$ axis is $\ln A$ and the gradient of the line is $-\frac{E_a}{R}$.

Worked example

In a reaction the values of k were determined at different temperatures. The units of k are s^{-1}. A graph of $\ln k$ against $1/T$ was drawn (Figure 16).

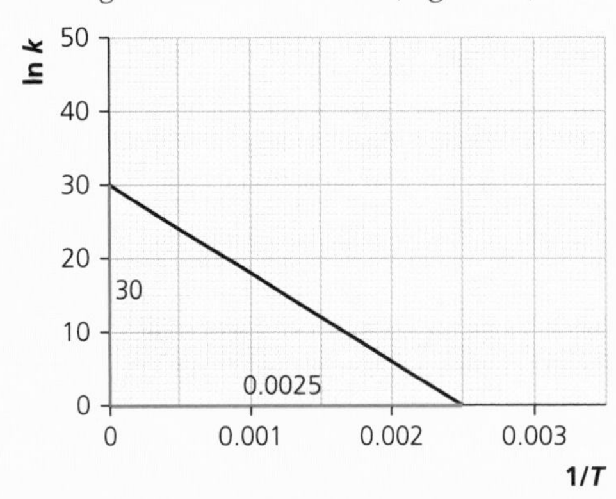

Figure 16

From this graph, the intercept on the $\ln k$ axis is 30. This means that $\ln A$ is 30.

So $A = e^{30} = 1.07 \times 10^{13}\,s^{-1}$.

The units of A are the same as the units of k.

The gradient is $-\frac{E_a}{R}$. The calculated gradient $= -\frac{30}{0.0025} = -12000$.

$$\frac{E_a}{R} = 12\,000\ (R = 8.31\,J\,K^{-1}\,mol^{-1})$$

$$E_a = 12\,000 \times 8.31 = 99\,720\,J\,mol^{-1}$$

$$E_a = 99.7\,kJ\,mol^{-1}$$

Knowledge check 8

What are the units of E_a in the Arrhenius equation?

Order linked to mechanism

The rate determining step in the mechanism for a reaction is the slowest step in the reaction.

Reactants which do not take part in the rate determining step have zero order.

Reactants which *do* take part in the rate determining step have order of 1 or 2.

Worked example

The reaction between X and Y is shown overall by the equation $X + 2Y \rightarrow 3Z$.

The rate equation for the reaction is:

$$\text{rate} = k[X]^2$$

Which one of the following is a suitable mechanism for the reaction?

A $X \rightarrow Y$ (slow step) $3Y \rightarrow 3Z$ (fast step)

B $X + Y \rightarrow A$ (slow step) $A + Y \rightarrow 3Z$ (fast step)

C $2Y \rightarrow B$ (slow step) $B + X \rightarrow 3Z$ (fast step)

D $X + Y \rightarrow$ (slow step) $Y + Z \rightarrow 4Z$ (fast step)

Answer

As X is the only reactant in the overall equation that is not zero order, only X can be a reactant in the rate determining step, which is the slowest step in the mechanism. The answer is A, as all other slow steps include Y as a reactant.

Required practical 7

Measuring the rate of reaction

Students will be required to measure the rate of reaction by an initial method such as measuring gas volume or by a continuous rate method such as titration or colorimetry. To carry out this practical you must be familiar with the content on pages 25 to 29.

Summary

- The rate law states that the rate is proportional to the concentration of the reactants raised to the power of the orders of reaction with respect to each reactant.
- The rate constant is k and it links the rate of reaction to the concentration of the reactants in the rate equation.
- The order of reaction with respect to a specific reactant is determined experimentally by measuring the rate of reaction against known concentrations of this reactant.
- The overall order of reaction is the sum of the individual order of reactions.
- Zero-order reactants do not take part in the rate determining step.
- The Arrhenius equation links the rate constant (k) with the activation energy (E_a) and temperature (T).
- A graph of $\ln k$ against $1/T$ gives a straight line with a gradient of $-E_a/R$ and an intercept on the $\ln k$ axis of $\ln A$.

Equilibrium constant K_p for homogeneous systems

K_p is an equilibrium constant calculated from partial pressure of reactants and products (measured in pressure units) for homogeneous gaseous reactions.

All equilibrium constants are only constant at constant temperature. The temperature should be quoted when the value of any equilibrium constant is given. If temperature remains constant, the equilibrium constant will not change. If any other factor is varied, such as pressure or concentration of reactants, the value of K_p remains constant.

Exam tip

K_p can be determined for homogeneous gaseous reactions since the concentration of a gas can be calculated as the number of moles of the gas in a certain volume (in dm^3).

K_p expressions and units of K_p

For the reaction:

$$aA(g) + bB(g) \rightleftharpoons cC(g) + dD(g)$$

$$K_p = \frac{(pC)^c(pD)^d}{(pA)^a(pB)^b}$$

where pC represents the partial pressure of C in the equilibrium mixture and c is the balancing number for C in the equation for the reaction. The same applies to A, B and D. The partial pressures of all products are on the top line of the expression raised to the power of their balancing numbers and the partial pressures of all reactants are on the bottom line again raised to the power of their balancing numbers.

$$\text{units of } K_p = \frac{(\text{pressure unit})^{(c+d)}}{(\text{pressure unit})^{(a+b)}}$$

The units of K_p are expressed in the same units as the units of the total pressure. If $c + d = a + b$, then K_p will have no units.

Exam tip

The position of equilibrium may vary when external factors are changed, but only changes in temperature will affect the value of the equilibrium constant. This is a common question.

Worked example

Write an expression for K_p for the reaction:

$$N_2(g) + 3H_2(g) \rightleftharpoons 2NH_3(g)$$

and calculate its units when the total pressure is measured in Pa.

Answer

$$K_p = \frac{(pNH_3)^2}{(pN_2)(pH_2)^3}$$

$$\text{units of } K_p = \frac{(Pa)^2}{(Pa)^4} = Pa^{-2}$$

Knowledge check 9

Write a K_p expression for the reaction:

$N_2O_4(g) \rightleftharpoons 2NO_2(g)$

Mole fractions and partial pressures

The mole fraction is the number of moles of that particular gas at equilibrium divided by the total number of moles of all gases at equilibrium.

$$\text{mole fraction} = \frac{\text{number of moles of particular gas at equilibrium}}{\text{total number of moles of gas at equilibrium}}$$

The partial pressure is the mole fraction of a gas in the equilibrium mixture multiplied by the total pressure of the system.

partial pressure = mole fraction × total pressure

The units of the partial pressure are the same as units of the overall pressure.

For homogeneous gaseous reactions partial pressures may be used to calculate a value for K_p.

Where P is the overall pressure, partial pressures are calculated as shown from the mole fraction.

A lower case p followed by the substance is used to represent the partial pressure. For example, the partial pressure of A is pA; the partial pressure of N_2 is pN_2.

For the reaction $A + B \rightleftharpoons C + D$:

$$K_p = \frac{(pC)(pD)}{(pA)(pB)}$$

$$\text{units} = \frac{(\text{pressure})^2}{(\text{pressure})^2} = \text{no units}$$

Calculating K_p and using K_p

The calculations involving K_p may involve calculating K_p or using K_p to calculate an equilibrium number of moles, a partial pressure, the total pressure on a system or an initial number of moles (Table 5).

Table 5 Blank table for calculations involving K_p

$$A + B \rightleftharpoons C + D$$

Initial moles					← Put initial values in here
Reacting moles					← These must be in ratio of balancing number in equation
Equilibrium moles					← Put final mole values in here
Mole fraction					← Calculated from total moles
Partial pressure					← mole fraction × overall pressure

If K_p is given, then the above expression table is filled in using x as the number of moles of x which react (for example, $-x$ in the reacting moles box for A as explained in the section on using K_c at AS). If K_p is given, the equilibrium moles may be in terms of x when K_p has no units. If you are expected to calculate the total pressure, it can be

Exam tip

The total of all the mole fractions should add up to 1. The total of the partial pressures should add up to the total pressure. This is a handy check if you are asked to calculate the mole fractions and then the partial pressures.

Exam tip

Again remember that the total of the partial pressures must add to the total pressure. In this general example: $pA + pB + pC + pD = P$

Exam tip

Again, remember that if K_p has no units, then K_p can be calculated from mole fractions instead of partial pressures.

Exam tip

Initial masses of gases present or even masses of gases present at equilibrium may be given. Simply divide the mass of these gases by the M_r of the gas to determine the initial moles or equilibrium moles and fill them into the correct position in the table. Watch out for the diatomic gases because you have to use the M_r of the diatomic gas: for example, 32 for O_2.

written as P and the partial pressures will be in terms of P. These can be substituted into the K_p expression and solved for P.

Calculating K_p

Worked example

Calculate a value for K_p for the equilibrium:

$$2SO_2(g) + O_2(g) \rightleftharpoons 2SO_3(g)$$

if an original mixture of 16.0 g of sulfur dioxide and 4.0 g of oxygen yields 16.0 g of sulfur trioxide at equilibrium, at a total pressure of 1.1×10^6 Pa at 450°C.

Answer

$$K_p = \frac{(pSO_3)^2}{(pSO_2)^2(pO_2)}$$

$$\text{units of } K_p = \frac{(Pa)^2}{(Pa)^2(Pa)} = \frac{1}{(Pa)} = Pa^{-1}$$

From the masses:

$$\text{Initial moles of } SO_2 = \frac{16.0}{64.1} = 0.250 \text{ mol}$$

$$\text{Initial moles of } O_2 = \frac{4.00}{32.0} = 0.125 \text{ mol}$$

$$\text{Equilibrium moles of } SO_3 = \frac{16.0}{80.1} = 0.200 \text{ mol}$$

Completing the table with the given data:

$$2SO_2 + O_2 \rightleftharpoons 2SO_3$$

	$2SO_2$	O_2	$2SO_3$
Initial moles	0.250	0.125	0
Reacting moles			
Equilibrium moles			0.200
Mole fraction			
Partial pressure			

If 0.200 moles of SO_3 are present at equilibrium, then 0.200 moles of SO_2 and 0.100 moles of O_2 must have reacted to form 0.200 moles of SO_3.

→

The table now looks like the one below:

$$2SO_2 + O_2 \rightleftharpoons 2SO_3$$

Initial moles	0.250	0.125	0
Reacting moles	−0.200	−0.100	+0.200
Equilibrium moles	0.050	0.025	0.200
Mole fraction			
Partial pressure			

The mole fraction is calculated from the total equilibrium moles of gas = 0.050 + 0.025 + 0.200 = 0.275.

The moles fractions are calculated as shown in the table below:

$$2SO_2 + O_2 \rightleftharpoons 2SO_3$$

Initial moles	0.250	0.125	0
Reacting moles	−0.200	−0.100	+0.200
Equilibrium moles	0.050	0.025	0.200
Mole fraction	$\frac{0.050}{0.275}$	$\frac{0.025}{0.275}$	$\frac{0.200}{0.275}$
Partial pressure	$\frac{0.050}{0.275} \times 1.1 \times 10^6$ = 2.00×10^5 Pa	$\frac{0.025}{0.275} \times 1.1 \times 10^6$ = 1.00×10^5 Pa	$\frac{0.200}{0.275} \times 1.1 \times 10^6$ = 8.00×10^5 Pa

$$K_p = \frac{(pSO_3)^2}{(pSO_2)^2(pO_2)} = \frac{(8.00 \times 10^5)^2}{(2.00 \times 10^5)^2 \times (1.00 \times 10^5)} = 1.60 \times 10^{-4}\,\text{Pa}^{-1}$$

Exam tip

This is the first type of equilibrium calculation, which has two numbers multiplied on the bottom line. As a rule for K_p expressions, always calculate the top line and bottom line separately and then divide, as this avoids common calculator errors. Unless you are very proficient with a calculator and the use of brackets you will make mistakes. Check the answer twice.

Knowledge check 10

What is the sum of all the mole fractions in a gaseous reaction?

Calculations using K_p

For the equilibrium $H_2(g) + I_2(g) \rightleftharpoons 2HI(g)$, $K_p = 2.00$ at 1500 K. Initially there is 1.00 mole of H_2 and 1.00 mole of I_2. Calculate the number of moles of HI present at equilibrium.

$$K_p = \frac{(pHI)^2}{(pH_2)(pI_2)}$$

$$\text{units of } K_p = \frac{(\text{pressure})^2}{(\text{pressure})^2} = \text{no units}$$

As there is 1.00 mole of H_2 and 1.00 mole of I_2 present initially, the table can be completed as shown up to the equilibrium moles.

	H_2 +	I_2 $\rightleftharpoons$	$2HI$
Initial moles	1.00	1.00	0
Reacting moles	$-x$	$-x$	$+2x$
Equilibrium moles	$1.00 - x$	$1.00 - x$	$2x$
Mole fraction			
Partial pressure			

The total equilibrium moles of gas are determined = $(1.00 - x) + (1.00 - x) + 2x = 2$. The total equilibrium moles of gas are not given in terms of x and this is common in this style of question where K_p has no units. The total pressure is taken as P.

	H_2 +	I_2 $\rightleftharpoons$	$2HI$
Initial moles	1.00	1.00	0
Reacting moles	$-x$	$-x$	$+2x$
Equilibrium moles	$1.00 - x$	$1.00 - x$	$2x$
Mole fraction	$\frac{1.00 - x}{2}$	$\frac{1.00 - x}{2}$	$\frac{2x}{2}$
Partial pressure	$\frac{1.00 - x}{2} \times P$	$\frac{1.00 - x}{2} \times P$	$\frac{2x}{2} \times P$

The mole fractions can be used in place of the partial pressures in the K_p expression as there are no units of K_p. The total pressure, P, would cancel out in the equilibrium expression.

$$K_p = \frac{(pHI)^2}{(pH_2)(pI_2)} = \frac{\left(\frac{2x}{2}\right)^2}{\left(\frac{1.00-x}{2}\right)^2} = \frac{\frac{(2x)^2}{4}}{\frac{(1.00-x)^2}{4}} = \frac{(2x)^2}{4} \times \frac{4}{(1.00-x)^2}$$

$$= \frac{(2x)^2}{(1.00-x)^2} = \left(\frac{2x}{1.00-x}\right)^2$$

$K_p = 2.00$

So:

$$\left(\frac{2x}{1.00-x}\right)^2 = 2.00$$

Taking a square root of both sides:

$$\left(\frac{2x}{1-x}\right) = \sqrt{2.00}$$

So:

$$\frac{2x}{1-x} = 1.41$$

$$2x = 1.41(1-x) = 1.41 - 1.41x$$

$3.41x = 1.41$, which gives:

$$x = \frac{1.41}{3.41} = 0.414$$

Once the value of x is calculated it can be filled into the values in the table to determine mole fractions. If the total pressure was given, the partial pressure can be determined.

$$H_2 + I_2 \rightleftharpoons 2HI$$

	H_2	I_2	2HI
Initial moles	$1.00 - x$	$1.00 - x$	$2x$
When $x = 0.414$	0.586	0.586	0.828

These are the mole fractions. So the answer to the question is that the number of moles of HI present at equilibrium is 0.828.

Exam tip

When you have determined the value of x, always check the question to see exactly which quantity you are being asked to determine. It may be $1 - x$ or in this case it is $2x$. Many candidates think they have finished the question when they determine the value of x, but the answer may involve using x further.

Exam tip

Answers to these questions, and those in which we use K_p, can be checked, if time permits, by putting your answers into the expression for the equilibrium constant. For example, in the above example:

$$K_p = \frac{(0.828)^2}{(0.586)(0.586)} = 2.00$$

Exam tip

This seems complicated maths but it is basically fractions and indices and you will be become more proficient with practice. Remember that a fraction divided by another fraction is simply the top fraction multiplied by the bottom fraction upside down. And if both quantities in a fraction are squared (top and bottom) then it is a simple fraction all squared.

Calculating the total pressure

Often P is used for the total P where it has to be determined. You will be given enough information in the question to determine P from the K_p value or sometimes from a partial pressure.

Worked example

If sulfur dioxide and oxygen are mixed in a 2.00 moles SO_2 to 1.00 mole O_2 ratio and heated to 1050 K, they reach equilibrium according to the equation:

$$2SO_2(g) + O_2(g) \rightleftharpoons 2SO_3(g) \qquad K_p = 1.25 \times 10^{-4}\,Pa^{-1}$$

A 50% conversion of sulfur dioxide to sulfur trioxide occurred. Determine the total pressure on the system.

Answer

$$K_p = \frac{(pSO_3)^2}{(pSO_2)^2(pO_2)} = 1.25 \times 10^{-4}\,Pa^{-1}$$

	$2SO_2$	+ O_2	$\rightleftharpoons$ $2SO_3$
Initial moles	2.00	1.00	0
Reacting moles	−1.00	−0.500	+1.00
Equilibrium moles	1.00	0.500	1.00
Mole fraction	$\frac{1.00}{2.50} = 0.400$	$\frac{0.500}{2.50} = 0.200$	$\frac{1.00}{2.50} = 0.400$
Partial pressure	$0.400P$	$0.200P$	$0.400P$

If 50% of SO_2 is converted into SO_3 then 1.00 mole out of the 2.00 moles of SO_2 reacts

Total equilibrium moles = 2.50

$$K_p = \frac{(0.400P)^2}{(0.400P)^2(0.200P)} = \frac{1}{0.200P} = 1.25 \times 10^{-4}\,Pa^{-1}$$

So:

$$1 = 0.200P \times 1.25 \times 10^{-4} = 2.5 \times 10^{-5}P$$

$$P = \frac{1}{2.5 \times 10^{-5}} = 40000\,Pa$$

Exam tip

At this stage $(0.400P)^2$ cancels down to 1 in the expression, leaving $1/0.200P$. You can still multiply out the $(0.400P)^2$ to get $0.160P^2$, but remember that both the 0.400 and the P need to be squared.

Position of equilibrium and the equilibrium constant

The position of equilibrium and K_c were studied earlier in the course. The position of equilibrium may be affected by changes in temperature, pressure and concentration (this should be revised from AS), but only a change in temperature affects the value of K_c. The same applies to K_p.

Only a change in temperature will affect the value of K_p.

- For an exothermic forward reaction, an increase in temperature will decrease the value of K_p.
- For an endothermic reaction, an increase in temperature will increase the value of K_p.

This can be used to predict whether a reaction is endothermic or exothermic with given values of K_p.

Summary

- $K_p = \frac{(pC)^c(pD)^d}{(pA)^a(pB)^b}$ for the gaseous equilibrium $aA + bB \rightleftharpoons cC + dD$.
- K_p is constant at constant temperature; its value is not affected by changes in concentration, volume or pressure.
- pA represents the partial pressure of gas A in the mixture and is calculated from the mole fraction × total pressure.
- The mole fraction is calculated from the equilibrium moles of a substance divided by the total equilibrium moles.
- The sum of the mole fractions is 1.
- The sum of the partial pressures is the total pressure.
- The units of K_p are pressure units and there may be no units depending on *a*, *b*, *c* and *d*.
- For an equilibrium which has no units of K_p, the equilibrium moles or mole fractions may be used in the K_p expression to calculate a value for K_p.

Exam tip

The one to remember is that a change in pressure has *no effect* on the value of K_p. The value of K_p is only affected by changes in temperature.

Knowledge check 11

What effect, if any, does increasing the pressure have on the position of equilibrium and the value K_p for the reaction:

$N_2(g) + 3H_2(g) \rightleftharpoons 2NH_3(g)$ where $\Delta H = -92.2\,kJ\,mol^{-1}$?

Electrode potentials and electrochemical cells

When a metal is placed in a solution of its ions, an equilibrium is established between the ions in the solution and the metal atoms.

The equilibrium can be represented by a half-equation:

$$M^{n+}(aq) + ne^- \rightarrow M(s)$$

for example:

$$Zn^{2+}(aq) + 2e^- \rightarrow Zn(s)$$

$$Fe^{2+}(aq) + 2e^- \rightarrow Fe(s)$$

The ions are aqueous and the metal is solid.

- The half-equations are written as reduction reactions (gain of electrons).
- The metal dipping into a solution of its ions is called a half cell.
- Two half cells combined are called a cell.
- In one of the half cells an oxidation must occur and in the other a reduction must occur.
- In any electrochemical reaction, there must be a reduction and an oxidation.
- Often normal arrows are used to represent the reactions occurring in the half cell.

Standard electrode potentials

The **electrode potential** is a relative measure of the feasibility of the reduction occurring.

Electrode potentials are represented by $E^{\ominus}$ and are measured in volts (V).

The electrode potential must be stated with a sign and the units V.

For example:

$Fe^{2+}(aq) + 2e^- \rightarrow Fe(s) \quad E^{\ominus} = -0.44\,V$

$Zn^{2+}(aq) + 2e^- \rightarrow Zn(s) \quad E^{\ominus} = -0.76\,V$

A negative electrode potential indicates that the reduction reaction is not feasible unless a more feasible oxidation takes place.

The reduction of iron(II) ions to iron is not feasible, but if iron(II) ions were mixed with zinc metal, the zinc metal would oxidise:

$Zn(s) \rightarrow Zn^{2+}(aq) + 2e^- \quad +0.76\,V$

and the iron(II) would be reduced to iron:

$Fe^{2+}(aq) + 2e^- \rightarrow Fe(s) \quad -0.44\,V$

The **standard electrode potential** is the electrode potential of a standard electrode with ion concentration of 1.00 $mol\,dm^{-3}$ at 298 K connected to a standard hydrogen electrode (1.00 $mol\,dm^{-3}$ $H^+(aq)$, 298 K and 100 kPa $H_2(g)$) using a high resistance voltmeter and a salt bridge.

Exam tip

Standard electrode potentials are not measured per mole of anything, as they are all carried out under standard solution conditions. It does not matter how many moles of each reactant react.

Electromotive force (emf)

The **electromotive force (emf)** of a complete cell is the total value of the oxidation and reduction standard electrode potentials when the two half cells are combined.

- A positive emf indicates a feasible reaction.
- A negative emf indicates an unfeasible reaction.
- The emf always relates to the overall redox reaction (oxidation and reduction combined).

The overall electromotive force (emf) of this cell is +0.76 − 0.44 = +0.32 V.

So this overall reaction is feasible. A positive emf indicates that the reaction is feasible.

The $E^{\ominus}$ values given are actually standard electrode potentials. The standard means they have been compared to a standard, which is the standard hydrogen electrode.

The **electromotive force (emf)** is the potential difference measured when two half cells are combined.

Features of standard hydrogen electrode

Figure 17 shows the features of the standard hydrogen electrode:

- platinum electrode
- 1.00 mol dm^{-3} hydrogen ions in solution
- H_2 gas at 100 kPa pressure
- the temperature should be 298 K
- the standard electrode potential of the standard hydrogen electrode is 0.00 V

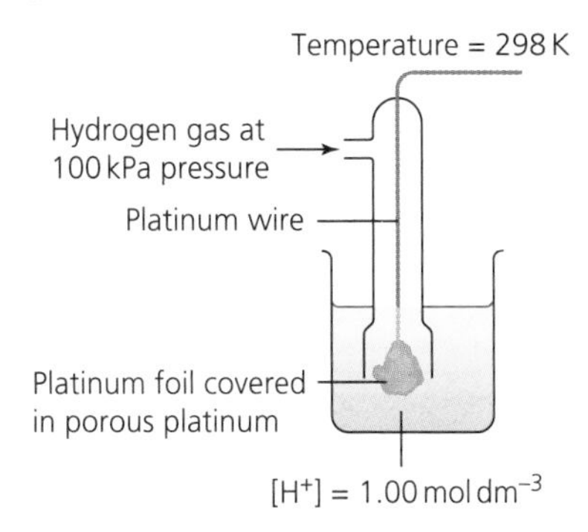

Figure 17 The standard hydrogen electrode

Knowledge check 12

What conditions are required for the standard hydrogen electrode?

Combination of half cells

Two half cells can be combined using a salt bridge and an external circuit containing a high resistance voltmeter to produce a cell.

Exam tip

A high resistance voltmeter is used to measure the emf of the cell, but does not allow current to flow in the circuit. Any flow of current would change the concentration of ions in the solutions and the cell would no longer be under standard conditions.

The salt bridge is usually filter paper soaked in saturated potassium nitrate solution which dips into both half cell solutions. The salt bridge allows electrical connection between the half cells without allowing them to mix.

- The left-hand half cell (by convention) is the one in which oxidation occurs. This is the negative electrode of the cell.
- The right-hand half cell (by convention) is the one in which reduction occurs. This is the positive electrode of the cell.

Figure 18 shows the combination of a standard hydrogen half cell or electrode on the left with a zinc half cell or electrode on the right.

Exam tip

Potassium chloride may be used in place of potassium nitrate for the salt bridge, but this may interfere with the cell, as Cl^- ions can form complexes with transition metal ions so it is to be avoided with cells containing these ions.

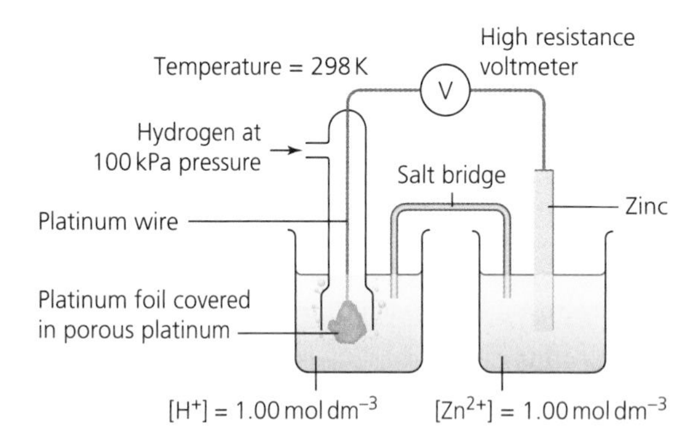

Figure 18 Standard hydrogen half cell and zinc half cell

Exam tip

A common question can be to sketch the right-hand cell attached to the standard hydrogen electrode. You would not usually be expected to draw the standard hydrogen electrode. Sketch the diagram from the voltmeter through the labelled right-hand cell to the salt bridge connecting to the standard hydrogen electrode.

Left-hand cell or right-hand cell

For the two standard electrode potentials given below:

$Zn^{2+}(aq) + 2e^- \rightarrow Zn(s) \quad E^\ominus = -0.76\,V$

$Cd^{2+}(aq) + 2e^- \rightarrow Cd(s) \quad E^\ominus = -0.40\,V$

There are several ways to approach working out which is the left-hand cell and which is the right-hand cell but the easiest idea to keep in your head is that the most negative one, or the one closest to negative if they are both positive, is the oxidation half cell and should be on the left. It will form the negative electrode.

Exam tip

NEGATOX is a good way of remembering which electrode is negative as the **negat**ive electrode is the one at which **ox**idation occurs.

So the zinc half cell is on the left and the cadmium half cell is on the right. The overall ionic equation for the reaction would be:

$Zn + Cd^{2+} \rightarrow Zn^{2+} + Cd$

Exam tip

You worked through balancing half-equations at AS. Simply reverse the more negative equation (change $Zn^{2+} + 2e^- \rightarrow Zn$ to $Zn \rightarrow Zn^{2+} + 2e^-$). The electrons are balanced ($2e^-$ in each equation) and combine them to cancel out the electrons.

The emf of this cell is found by adding together the $E^\ominus$ values with the sign of the left-hand one changed. In this example it is + 0.76 − 0.44 = + 0.32 V. The + 0.76 is $E^\ominus$ value for the zinc half cell with its sign changed. A diagram of the cell is shown in Figure 19.

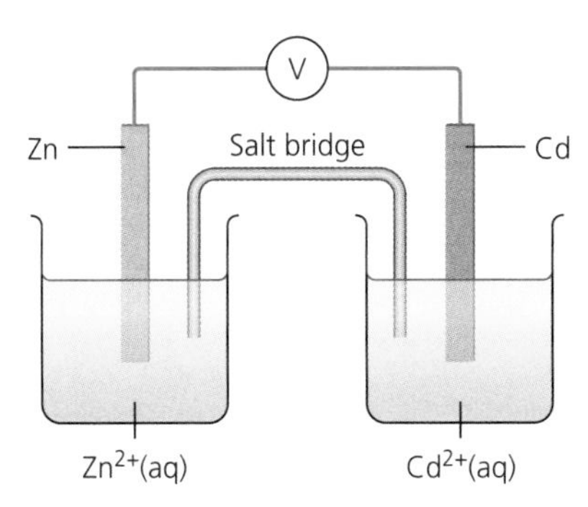

Figure 19 Zinc and cadmium cell

Non-standard conditions in a cell

Two identical cells or two cells with the same electrode potential will show no overall voltage. However, two identical cells with different concentration of solution will show a reduction where the ion in solution is at a greater concentration.

The left-hand cell in Figure 20 will show a reduction ($Cu^{2+} + 2e^- \rightarrow Cu$), as the Cu^{2+} ions are at higher concentration.

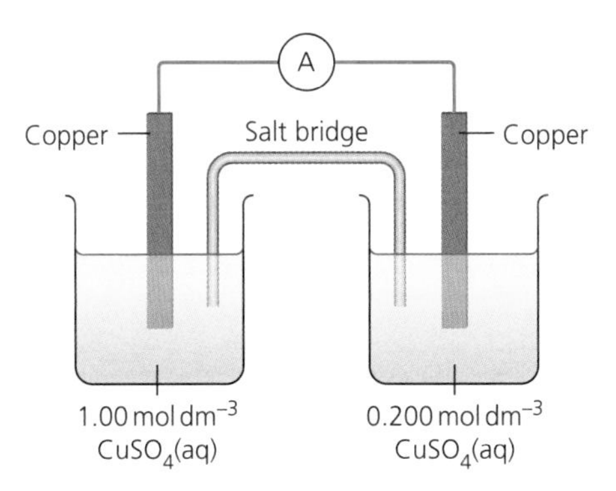

Figure 20 Two copper half cells

Note that the ammeter between the cells allows current to flow. A voltmeter is not used here as you are not trying to measure the emf of the cell.

The mass of the left-hand electrode will increase as copper is deposited on it and the mass of the right-hand one will decreases as $Cu \rightarrow Cu^{2+} + 2e^-$, so copper in the electrode is being used up.

Conventional cell representation

Cells are written as oxidation at left-hand half cell || reduction at right-hand half cell.

State symbols are not absolutely necessary in cell notation but should be included if asked for in a question.

For example:

$Zn(s) \mid Zn^{2+}(aq) \parallel Cu^{2+}(aq) \mid Cu(s)$

|| represents the salt bridge.

$Zn(s) \mid Zn^{2+}(aq)$ represents the oxidation reaction $Zn \rightarrow Zn^{2+} + 2e^-$

Exam tip

Indicating the direction of flow of electrons can be a common question. The electrons are flowing from the left-hand side (as they are given up by $Zn \rightarrow Zn^{2+} + 2e^-$) to the right-hand side (where they are used as $Cd^{2+} + 2e^- \rightarrow Cd$).

$Cu^{2+}(aq) | Cu(s)$ represents the reduction reaction $Cu^{2+} + 2e^- \rightarrow Cu$

The overall emf of the cell can be calculated using the expression: $E_{rhs}^\ominus - E_{lhs}^\ominus$ (right-hand side – left-hand side).

For $Zn^{2+} + 2e^- \rightarrow Zn$: $E^\ominus = -0.76\,V$

For $Cu^{2+} + 2e^- \rightarrow Cu$: $E^\ominus = +0.34\,V$

Therefore for the cell $Zn(s) | Zn^{2+}(aq) || Cu^{2+}(aq) | Cu(s)$.

emf = +0.34 – (–0.76) = +1.10 V

This can be written more simply without state symbols as: $Zn | Zn^{2+} || Cu^{2+} | Cu$. This is common.

This reaction is feasible as the overall emf is positive.

Exam tip

This is a fast way of doing it from cell notations but again the sign of the oxidation side is changed, hence $-E_{lhs}^\ominus$. The – changes the sign.

Phase boundaries

The | used in cell representation is called a phase boundary and should be used when the two species in a half cell are in different phases or states. For example, $Zn(s) | Zn^{2+}(aq)$.

However, if the two species are in the same phase a comma should be used between them. For example, $Fe^{3+}(aq), Fe^{2+}(aq)$.

Pt should be included if it is used as the electrode for gas electrodes and also for those with two ions in solution.

Worked example

The two half cells

$Fe^{3+}(aq) + e^- \rightarrow Fe^{2+}(aq)$ $E^\ominus = +0.77\,V$

$Cl_2(g) + 2e^- \rightarrow 2Cl^-(aq)$ $E^\ominus = +1.36\,V$

Write the equation for the overall cell reaction. Give the conventional cell representation for this cell, identify the positive electrode and calculate the emf of the cell.

Answer

First, the more negative one is the one that will be reversed. Both have positive $E^\ominus$, but the more negative is the Fe half cell as it is closer to being negative. This will be the oxidation half cell and will be on the left.

The overall equation requires the Fe half cell to be reversed and multiplied by 2 to balance the electrons.

$2Fe^{2+} \rightarrow 2Fe^{3+} + 2e^-$

$Cl_2 + 2e^- \rightarrow 2Cl^-$

The overall equation is: $2Fe^{2+} + Cl_2 \rightarrow 2Fe^{3+} + 2Cl^-$

→

The cell representation is: Pt | Fe^{2+}, Fe^{3+} || Cl_2 | Cl^- | Pt

- Note that Pt has to be included at either end as it is used as the electrode in both half cells and it should be separated by a |.
- The comma between Fe^{2+} and Fe^{3+} is there as they are both in solution, both (aq), so in the same phase.
- $2Cl^-$ is not needed as it is just the species that is needed.
- A | is required between Cl_2 and Cl^- as Cl_2 is a gas and Cl^- ions are in solution.

Using NEGATOX, the negative electrode is the one at which oxidation occurs. Oxidation occurs at the iron electrode, so it is negative, which means that the chlorine electrode is the positive electrode.

The emf is +1.36 – (–0.77) = +2.13 V.

Feasibility of reactions

A positive emf for an electrochemical cell is an indication that the reaction between the species would occur under standard conditions. This can be used to predict the feasibility of normal chemical reactions.

From the following standard electrode potentials:

$Zn^{2+} + 2e^- \rightarrow Zn \quad E^\ominus = -0.76\,V$

$Cu^{2+} + 2e^- \rightarrow Cu \quad E^\ominus = +0.34\,V$

$Mg^{2+} + 2e^- \rightarrow Mg \quad E^\ominus = -2.37\,V$

$Fe^{2+} + 2e^- \rightarrow Fe \quad E^\ominus = -0.44\,V$

Consider the reaction between zinc and iron(II) ions in solution. The ionic equation for the reaction is:

$Zn + Fe^{2+} \rightarrow Zn^{2+} + Fe$

When a piece of zinc is placed in a solution containing iron(II) ions, a displacement reaction occurs. From GCSE we would explain that this is due to zinc being more reactive than iron, but it is more accurately explained using the electrode potentials. A positive emf for the electrochemical cell containing these two half cells indicates a feasible reaction. The answers are usually presented in one of two ways:

1. Explain that a reaction occurs by considering that the $E^\ominus$ value for the reduction of Fe^{2+} to Fe is greater than the $E^\ominus$ value for the reduction of Zn^{2+} to Zn. This would mean that the reduction of Fe^{2+} to Fe is more favourable, so Zn reacts with Fe^{2+} ions in solution. This is often simply written in mark schemes as $E^\ominus (Fe^{2+} | Fe) > E^\ominus (Zn^{2+} | Zn)$.
2. It can also be explained by calculating the emf of the cell.

$Zn^{2+} + 2e^- \rightarrow Zn \quad E^\ominus = -0.76\,V$

$Fe^{2+} + 2e^- \rightarrow Fe \quad E^\ominus = -0.44\,V$

The more negative one changes direction and becomes the oxidation. So:

$Zn \rightarrow Zn^{2+} + 2e^-$ +0.76 V

$Fe^{2+} + 2e^- \rightarrow Fe$ −0.44 V

emf = +0.32 V

The positive emf indicates that this is a feasible reaction. So zinc will reduce iron(II) ions to iron.

Link to reactivity

Extending the series we had above:

$Zn^{2+} + 2e^- \rightarrow Zn$	$E^\ominus = -0.76\,V$
$Cu^{2+} + 2e^- \rightarrow Cu$	$E^\ominus = +0.34\,V$
$Mg^{2+} + 2e^- \rightarrow Mg$	$E^\ominus = -2.37\,V$
$Fe^{2+} + 2e^- \rightarrow Fe$	$E^\ominus = -0.44\,V$
$I_2 + 2e^- \rightarrow 2I^-$	$E^\ominus = +0.54\,V$
$Br_2 + 2e^- \rightarrow 2Br^-$	$E^\ominus = +1.07\,V$
$Cl_2 + 2e^- \rightarrow 2Cl^-$	$E^\ominus = +1.36\,V$

The electrode potential for the reverse oxidation reaction gives a measure of the reactivity of the metal, as the most reactive metal will form its ions most easily.

From these values, it is clear that it is most feasible for Mg to form its ions (+2.37 V) followed by Zn (+0.76 V), followed by Fe (+0.44 V) and least of all Cu (−0.44 V).

This explains the order of reactivity of the metals.

The non-metals are reduced and it is most feasible for chlorine to be reduced (+1.36 V) compared to bromine (+1.07 V) and then iodine (+0.54 V). This explains why the reactivity of the halogens decreases down group 7.

Oxidising agents and reducing agents

Oxidising agents are easily reduced and reducing agents are easily oxidised.

	$E^\ominus$/V
$Mg^{2+} + 2e^- \rightarrow Mg$	−2.37
$Zn^{2+} + 2e^- \rightarrow Zn$	−0.76
$Fe^{2+} + 2e^- \rightarrow Fe$	−0.44
$Cu^{2+} + 2e^- \rightarrow Cu$	+0.34
$I_2 + 2e^- \rightarrow 2I^-$	+0.54
$Cl_2 + 2e^- \rightarrow 2Cl^-$	+1.36

The best reducing agent in the above set of half-equations is magnesium, as it is most easily oxidised ($Mg \rightarrow Mg^{2+} + 2e^-$ is + 2.37 V). The best oxidising agent is chlorine ($Cl_2 + 2e^- \rightarrow 2Cl^-$ is + 1.36 V).

Knowledge check 13

In the cell $Zn \mid Zn^{2+} \parallel Cu^{2+} \mid Cu$ which is the negative electrode?

Exam tip

Look to the left of the half-equations for the oxidising agents, as these species will gain electrons and so cause an oxidation to occur as some other species loses electrons. Look to the right of the half-equations for the reducing agents, as these species will lose electrons and so cause an oxidation to occur as some other species gains electrons.

Electrode potentials applied to vanadium chemistry

Transition metals have variable oxidation states and this will be studied in the next book. However, electrode potentials may be used to explain the oxidation and reduction processes. Vanadium compounds exist with vanadium in the +5, +4, +3 and +2 oxidation states. The half-equations below show the sequential reduction of vanadium.

Oxidation states	Half-equation	Electrode potential
+5 to +4	$VO_2^+ + 2H^+ + e^- \rightarrow VO^{2+} + H_2O$	$E^\ominus = +1.00\,V$
+4 to +3	$VO^{2+} + 2H^+ + e^- \rightarrow V^{3+} + H_2O$	$E^\ominus = +0.32\,V$
+3 to +2	$V^{3+} + e^- \rightarrow V^{2+}$	$E^\ominus = -0.26\,V$

Worked example

Which reducing agent will reduce vanadium from the +5 oxidation state to the +3 oxidation state but not to the +2 oxidation state?

A iodine
B iron
C sulfur dioxide
D zinc

given the following standard redox potentials:

	$E^\ominus$/V
$Zn^{2+}(aq) + 2e^- \rightarrow Zn(s)$	−0.76
$Fe^{2+}(aq) + 2e^- \rightarrow Fe(s)$	−0.44
$SO_4^{2-}(aq) + 2H^+(aq) + 2e^- \rightarrow 2H_2O(l) + SO_2(g)$	+0.17
$I_2(aq) + 2e^- \rightarrow 2I^-(aq)$	+0.54

Answer

The choice has to be from the right of these half-equations, as a substance which is to be oxidised (i.e. lose electrons) must be chosen. If any substance is given which is on the left of one of these equations (for example, iodine) then it can be ruled out, as it would be reduced.

Each reaction is taken in turn with each stage of the reduction of vanadium.

Try the oxidation of Fe to Fe^{2+}, which has a potential of +0.44 V:

$VO_2^+ + 2H^+ + e^- \rightarrow VO^{2+} + H_2O$	+1.00 V	
$Fe^{2+} + 2e^- \rightarrow Fe$	−0.44 V	
(Reverse second equation)	emf = +1.44 V	feasible

→

or $E^\ominus$ (VO_2^+/VO^{2+}) > $E^\ominus$ (Fe^{2+}/Fe), so vanadium is reduced from +5 to +4 by the oxidation of iron:

$VO^{2+} + 2H^+ + e^- \rightarrow V^{3+} + H_2O$	+0.32 V	
$Fe^{2+} + 2e^- \rightarrow Fe$	−0.44 V	
(Reverse second equation)	emf = +0.76 V	feasible

or $E^\ominus$ (VO^{2+}/V^{3+}) > $E^\ominus$ (Fe^{2+}/Fe), so vanadium is reduced from +4 to +3 by the oxidation of iron:

$V^{3+} + e^- \rightarrow V^{2+}$	−0.26 V	
$Fe^{2+} + 2e^- \rightarrow Fe$	−0.44 V	
(Reverse second equation)	emf = +0.18 V	feasible

or $E^\ominus$ (V^{3+}/V^{2+}) > $E^\ominus$ (Fe^{2+}/Fe), so vanadium is reduced from +3 to +2 by the oxidation of iron.

Fe cannot be the answer because it will reduce vanadium from +5 to +2.

Try the oxidation of sulfur dioxide to sulfate:

$VO_2^+ + 2H^+ + e^- \rightarrow VO^{2+} + H_2O$	+1.00 V	
$SO_4^{2-} + 4H^+ + 2e^- \rightarrow SO_2 + 2H_2O$	+0.17 V	
(Reverse second equation)	emf = +0.83 V	feasible

or $E^\ominus$ (VO_2^+/VO^{2+}) > $E^\ominus$ (SO_4^{2-}/SO_2), so vanadium is reduced from +5 to +4 by the oxidation of sulfur dioxide:

$VO^{2+} + 2H^+ + e^- \rightarrow V^{3+} + H_2O$	+0.32 V	
$SO_4^{2-} + 4H^+ + 2e^- \rightarrow SO_2 + 2H_2O$	+0.17 V	
(Reverse second equation)	emf = +0.15 V	feasible

or $E^\ominus$ (VO^{2+}/V^{3+}) > $E^\ominus$ (SO_4^{2-}/SO_2), so vanadium is reduced from +4 to +3 by the oxidation of sulfur dioxide:

$V^{3+} + e^- \rightarrow V^{2+}$	−0.26 V	
$SO_4^{2-} + 4H^+ + 2e^- \rightarrow SO_2 + 2H_2O$	+0.17 V	
(Reverse second equation)	emf = −0.43 V	**not** feasible

or $E^\ominus$ (V^{3+}/V^{2+}) < $E^\ominus$ (SO_4^{2-}/SO_2), so vanadium is **not** reduced from +3 to +2 by the oxidation of sulfur dioxide.

Sulfur dioxide (SO_2) is the answer, as it will reduce vanadium from +5 to +3, but will not reduce vanadium from +3 to +2.

Zinc will reduce vanadium from +5 to +2 and iodide ions (which were not given in the question) will reduce vanadium from +5 to +4 but no further.

Required practical 8

Measuring the emf of an electrochemical cell

Students should be able to measure the emf of a simple cell, such as Zn | Zn^{2+} || Cu^{2+} | Cu, using a high resistance voltmeter.

Commercial application of cells

The emf of an electrochemical cell can be used as a commercial source of electrical energy. These are commonly called cells or batteries. There are three main types:

1. Primary cells — these are single use and non-rechargeable.
2. Secondary cells — these can be recharged.
3. Fuel cells — these continuously produce an electrical current as long as they are supplied with a fuel.

Basic electrochemistry of cells

A cell has a positive electrode (labelled +) and a negative electrode (labelled –).

The electrons flow from the negative electrode to the positive electrode when the cell is being used.

At the negative electrode, an oxidation reaction occurs; electrons are lost and oxidation number increases.

At the positive electrode, a reduction reaction occurs; electrons are gained and oxidation number decreases.

Exam tip

Remember NEGATOX as this still applies. The **negat**ive electrode is where an **ox**idation reaction is occurring.

Primary cells

A typical primary cell is shown in Figure 21.

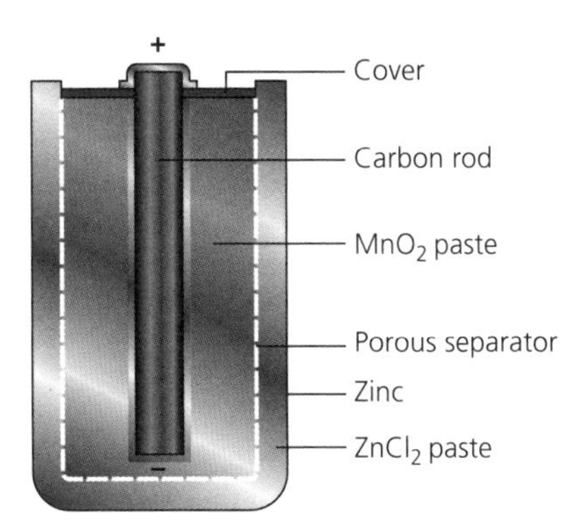

Figure 21 A typical non-rechargeable cell

At the negative electrode (oxidation occurs):

$$Zn \rightarrow Zn^{2+} + 2e^-$$

At the positive electrode (reduction occurs):

$$MnO_2 + H_2O + e^- \rightarrow MnO(OH) + OH^-$$

The half-equation for the reaction at the positive electrode is more unusual and should be learned. These types of reactions are common and you should copy them out until you know them and understand the changes in oxidation state.

In this half-equation, manganese is reduced from +4 to +3.

+4 in MnO_2 and +3 in $MnO(OH)$ as the O is −2 and the OH^- is overall −1, so Mn is +3.

You should also be able to combine half-equations to write an overall ionic equation for the reaction in the cell. The overall equation for this cell is:

$$Zn + 2MnO_2 + 2H_2O \rightarrow Zn^{2+} + 2MnO(OH) + 2OH^-$$

This reaction is non-reversible, so the cell, once discharged, cannot be used again.

Knowledge check 14

What is the oxidation state of manganese in MnO(OH)?

Exam tip

This was worked out by multiplying the reduction equation by 2 to balance electrons and adding them together.

Secondary cells

There are several types of secondary cells.

Nickel cadmium cell

At the negative electrode (oxidation occurs):

$$Cd + 2OH^- \rightarrow Cd(OH)_2 + 2e^-$$

At the positive electrode (reduction occurs):

$$NiO(OH) + H_2O + e^- \rightarrow Ni(OH)_2 + OH^-$$

The overall equation is:

$$Cd + 2NiO(OH) + 2H_2O \rightarrow Cd(OH)_2 + 2Ni(OH)_2$$

When any secondary cell is recharged, the equation above is simply reversed.

Sometimes the standard electrode potentials may be given and this allows the emf to be calculated, which is the voltage delivered by the cell or battery.

$$Cd(OH)_2 + 2e^- \rightarrow Cd + 2OH^- \qquad E^\ominus = -0.88\,V$$

$$NiO(OH) + H_2O + e^- \rightarrow Ni(OH)_2 + OH^- \qquad E^\ominus = +0.52\,V$$

The top half-equation is reversed in the cell so the emf = +0.52 + 0.88 = +1.40 V

Knowledge check 15

Explain in terms of oxidation states how nickel is reduced in this cell.

Lead–acid cell

This is the type of cell used in cars.

At the negative electrode (oxidation occurs):

$$Pb + HSO_4^- \rightarrow PbSO_4 + H^+ + 2e^-$$

At the positive electrode (reduction occurs):

$$PbO_2 + 3H^+ + HSO_4^- + 2e^- \rightarrow PbSO_4 + 2H_2O$$

The overall equation is:

$$PbO_2 + 2H^+ + 2HSO_4^- + Pb \rightarrow 2PbSO_4 + 2H_2O$$

Lead(II) sulfate is insoluble, so if the cell is not recharged for a long period of time the lead(II) sulfate builds up and the cell cannot be recharged. Again this equation is reversed as the cell is recharged.

Lithium ion cell

Lithium ion cells are the most commonly used cells in portable electronic devices such as mobile phones, tablets and laptops.

At the negative electrode (oxidation occurs):

$$Li \rightarrow Li^+ + e^-$$

At the positive electrode (reduction occurs):

$$Li^+ + CoO_2 + e^- \rightarrow Li^+[CoO_2]^-$$

The overall equation is:

$$Li + CoO_2 \rightarrow Li^+[CoO_2]^-$$

Graphite powder is used for the support medium in this type of cell, as water would react with lithium. Again the reaction is reversed when the cell is recharged.

The conventional cell representation for this cell is: $Li \mid Li^+ \parallel Li^+, CoO_2 \mid LiCoO_2 \mid Pt$.

Lithium is oxidised from 0 to +1 and cobalt is reduced from +4 in CoO_2 to +3 in $LiCoO_2$.

Fuel cells

Fuel cells convert chemical energy into electrical energy. There are several types of fuel cells.

Hydrogen fuel cell

A hydrogen fuel cell operates in acidic or alkaline conditions. The alkaline one is considered here.

At the negative electrode (oxidation occurs):

$$H_2 + 2OH^- \rightarrow 2H_2O + 2e^-$$

At the positive electrode (reduction occurs):

$$O_2 + 2H_2O + 4e^- \rightarrow 4OH^-$$

The overall equation is:

$$2H_2 + O_2 \rightarrow 2H_2O$$

The conventional cell representation for an alkaline hydrogen fuel cell is:

$$Pt \mid H_2 \mid OH^-, H_2O \parallel O_2 \mid H_2O, OH^- \mid Pt$$

Platinum is used for the electrodes. All species undergoing any redox change are included.

The overall equation in the hydrogen fuel cell is always the same and the emf is always +1.23 V, whether it is in alkaline or acidic conditions.

The current from a fuel cell should remain constant as long as the fuel is continuously supplied. The current from a primary or secondary cell will fall over time.

Risks and benefits

Primary cells generate waste and although some parts may be recycled they are generally just thrown away. Secondary cells can be recharged many times, but again they have a lifetime. Fuel cells continuously produce current as long as the fuel is supplied. Cells allow us to have portable electrical devices away from main electricity. However, they do contain metals, which are in ever-decreasing supply and should be recycled. Hydrogen fuel cells require a steady supply of hydrogen gas, which is flammable and so difficult to contain safely. Ethanol fuel cells are a possible alternative, as the ethanol may be produced by fermentation and so can be considered a carbon-neutral fuel since the CO_2 released by combustion is offset by the CO_2 taken in by photosynthesis.

Exam tip

In acidic conditions the oxidation reaction is $H_2 \rightarrow 2H^+ + 2e^-$, the reduction equation is $O_2 + 4H^+ + 4e^- \rightarrow 2H_2O$. The overall emf is +1.23 V and as $H_2 \rightarrow 2H^+ + 2e^-$ has $E^\ominus = 0.00$ V by definition, the $E^\ominus$ for the reduction reaction is +1.23 V.

Summary

- Standard electrode potentials give a measure in volts (V) of the feasibility of a half cell reaction.
- The emf (electromotive force) is the combination of the standard electrode potentials for the two half cell reactions.
- A positive emf indicates a feasible reaction, whereas a negative emf indicates a reaction which is not feasible.
- The standard hydrogen electrode is measured at 100 kPa pressure of H_2, 1 mol dm^{-3} of H^+ ions and 298 K.
- The strongest reducing agent (reductant) is the one which is most easily oxidised.
- The strongest oxidising agent (oxidant) is the one which is most easily reduced.
- A cell shows the reduction as the left-hand half cell and the reduction as the right-hand half cell.
- The negative electrode is where an oxidation reaction occurs (NEGATOX).
- Conventional cell representation shows the cell in terms of the oxidation on the left and the reduction on the right, using | as a phase boundary and || to show the salt bridge.
- Commercial electrochemical cells are either primary cells (non-rechargeable), secondary cells (rechargeable) or fuel cells.

Acids and bases

Brønsted–Lowry theory of acids and bases

The chemistry of acids and bases depends on protons. The terms proton and hydrogen ion are interchangeable, but proton is most often used.

In the reaction:

$$NH_3 + H_2O \rightarrow NH_4^+ + OH^-$$

- NH_3 accepts a proton to become NH_4^+. NH_3 acts as a **Brønsted–Lowry base**.
- H_2O donates a proton. H_2O acts as a **Brønsted–Lowry acid**.

A **Brønsted–Lowry base** is a proton acceptor.

A **Brønsted–Lowry acid** is a proton donor.

In the following reaction: $CH_3COOH + H_2O \rightarrow CH_3COO^- + H_3O^+$

- H_2O accepts a proton to become H_3O^+. H_2O acts as a Brønsted–Lowry base.
- CH_3COOH donates a proton. CH_3COOH acts as a Brønsted–Lowry acid.

Exam tip

Water can act as both a Brønsted–Lowry acid and base. H_3O^+ is the hydronium ion or hydroxonium ion or oxonium ion. It is the ion formed when acids react with water. All three names are acceptable.

Knowledge check 16

In the equation
$HCO_3^- + H_2O \rightarrow CO_3^{2-} + H_3O^+$
what is acting as a Brønsted–Lowry acid?

Classification of acids and bases

Acids (and bases) may be classified as strong or weak.

A **strong acid** is *completely* dissociated into its ions in solution.

Common strong acids are hydrochloric acid (HCl), sulfuric acid (H_2SO_4) and nitric acid (HNO_3). For strong acids, all of the acid molecules in solution are dissociated into ions.

For example: $HCl \rightarrow H^+ + Cl^-$

$H_2SO_4 \rightarrow 2H^+ + SO_4^{2-}$

$HNO_3 \rightarrow H^+ + NO_3^-$

HCl and HNO_3 are described as **monoprotic** acids. H_2SO_4 is a diprotic acid.

Monoprotic acids are acid molecules that donate one proton (H^+) per molecule of acid.

Exam tip

The equations for the dissociation of strong acids use a full arrow ($\rightarrow$) to show that the acid is fully dissociated. The dissociation equations can be written as the reaction of the acid, with water on the left forming the anion and H_3O^+. For example, $HCl + H_2O \rightarrow H_3O^+ + Cl^-$.

A **weak acid** is *slightly* dissociated into its ions in solution.

Common weak acids are carbonic acid (H_2CO_3), most carboxylic acids such as ethanoic acid (CH_3COOH), hydrocyanic acid (HCN), nitrous acid (HNO_2) and sulfurous acid (H_2SO_3).

For example:

$CH_3COOH \rightleftharpoons H^+ + CH_3COO^-$

$HNO_2 \rightleftharpoons H^+ + NO_2^-$

Generally this can be represented as:

$HA \rightleftharpoons H^+ + A^-$

where HA is the undissociated monoprotic weak acid.

Strong bases are completely dissociated into their ions in solution, whereas weak bases are slightly dissociated into their ions in solution.

Group 1 hydroxides are strong bases, for example, sodium hydroxide (NaOH) and potassium hydroxide (KOH). Ammonia is the most common example of a weak base.

Exam tip

Most of the weak acids you will encounter are monoprotic acids and any calculations will make that assumption. The reversible arrow ($\rightleftharpoons$) is used to show the slight dissociation. Again the dissociation equation can be written as the reaction of the acid with water forming the anion and H_3O^+. For example: $H_3COOH + H_2O \rightleftharpoons CH_3COO^- + H_3O^+$.

$NaOH \rightarrow Na^+ + OH^-$

$KOH \rightarrow K^+ + OH^-$

$NH_3 + H_2O \rightleftharpoons NH_4^+ + OH^-$

Soluble bases are called alkalis and they form hydroxide ions in solution.

Calculating pH

pH (always written with a small p and a capital H) is a logarithmic scale which gives a measure of the H^+ concentration in a solution.

$$pH = -\log_{10}[H^+]$$

where $[H^+]$ represents the concentration of H^+ ions in solution measured in $mol\,dm^{-3}$.

To calculate the hydrogen ion concentration from the pH, reverse the calculation

$$[H^+] = 10^{(-pH)}$$

A hydrogen ion concentration of $1.00 \times 10^{-2}\,mol\,dm^{-3}$ (= $0.0100\,mol\,dm^{-3}$) will give a pH of 2.00.

A solution with a pH of 8.00 has hydrogen ion concentration of $10^{-8.00}$ (= $1.00 \times 10^{-8}\,mol\,dm^{-3}$).

pH is usually quoted to 2 decimal places.

Knowledge check 17

Define pH.

pH of strong acids

Figure 22 links the concentration of the acid in $mol\,dm^{-3}$ shown as [acid] with the concentration of hydrogen ions shown as $[H^+]$ and the pH.

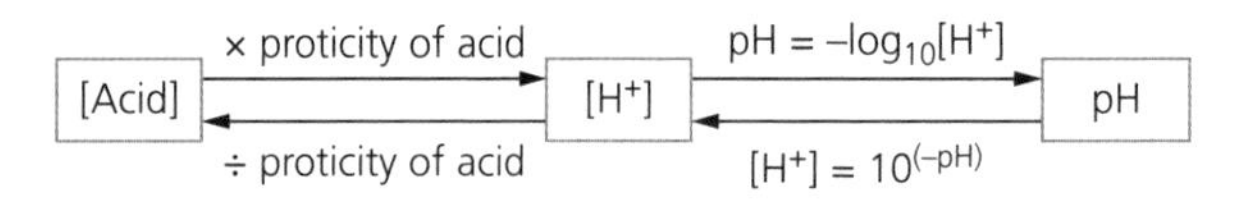

Figure 22 Determining pH of a strong acid

Worked example 1

Calculate the pH of $0.128\,mol\,dm^{-3}$ hydrochloric acid. Give your answer to 2 decimal places.

Answer

HCl is monoprotic, so $[H^+]$ = [acid].

$$[H^+] = 0.128\,mol\,dm^{-3}$$

$$pH = -\log_{10}[H^+] = -\log_{10}(0.128) = 0.89$$

Worked example 2

Calculate the pH of 1.05 mol dm^{-3} sulfuric acid. Give your answer to 2 decimal places.

Answer

H_2SO_4 is diprotic, so $[H^+] = 2 \times [H_2SO_4]$.

$$[H^+] = 2 \times 1.05 = 2.10\,\text{mol dm}^{-3}$$

$$pH = -\log_{10}[H^+] = -\log_{10}(2.10) = -0.32$$

Exam tip

When the concentration of hydrogen ions is above 1.00 mol dm^{-3}, the pH will be negative. If $[H^+]$ = 1.00 mol dm^{-3}, the pH will be 0. The above processes may be reversed. Check you can do these on your calculator.

Worked example 3

Calculate the concentration of a solution of sulfuric acid with a pH of 2.49. Give your answer to 3 significant figures.

Answer

$[H^+] = 10^{-2.49} = 0.00324\,\text{mol dm}^{-3}$

H_2SO_4 is diprotic, so $[H^+] = 2 \times [H_2SO_4]$

$$[H_2SO_4] = \frac{[H^+]}{2} = \frac{0.0324}{2} = 0.00162\,\text{mol dm}^{-3}$$

Exam tip

The same answer to 3 significant figures is obtained if the calculation is carried out completely on a calculator. However, as long as you show your working out for these questions, any rounding errors can be avoided. Work to 3 or 4 significant figures throughout the calculation, but give the answer to the required level of accuracy at the end.

Knowledge check 18

Calculate the pH of 2.54 mol dm^{-3} sulfuric acid.

Ionic product of water

Water slightly dissociates into hydrogen ions and hydroxide ions.

$$H_2O \rightleftharpoons H^+ + OH^-$$

- K_w is the ionic product of water and $K_w = [H^+][OH^-]$.
- Units of K_w are always mol^2 dm^{-6}.
- At 25.0°C, $K_w = 1.00 \times 10^{-14}$ mol^2 dm^{-6}.

Calculating the pH of water

In water $[H^+] = [OH^-]$, so $K_w = [H^+]^2$.

Worked example

At 50°C $K_w = 5.48 \times 10^{-14}\,mol^2\,dm^{-6}$. Calculate the pH of water at 40°C. Give your answer to 2 decimal places.

Answer

$$K_w = 5.48 \times 10^{-14} = [H^+]^2 \quad [H^+] = \sqrt{5.48 \times 10^{-14}} = 2.34 \times 10^{-7}\,mol\,dm^{-3}$$

$$pH = -\log_{10}[H^+] = -\log_{10}(2.34 \times 10^{-7}) = 6.63$$

Exam tip

We have become accustomed to think that the pH of pure water is 7, but this is only true at 25°C. As temperature increases above 25°C, the pH of water drops below 7. K_w increases as temperature increases, which indicates that the dissociation of water into H^+ and OH^- ions is endothermic, as the equilibrium is moving to the right (more H^+) as temperature increases.

pH of strong alkalis

Alkalis contain hydrogen ions, OH^-, in solution. If the concentration of OH^- ions is known, this can be converted to the concentration of H^+ ions using the K_w expression at a given temperature.

Figure 23 shows the connections between concentration of the alkali, [alkali], concentration of hydroxide ions, $[OH^-]$, concentration of hydrogen ions, $[H^+]$, and pH for a strong alkali.

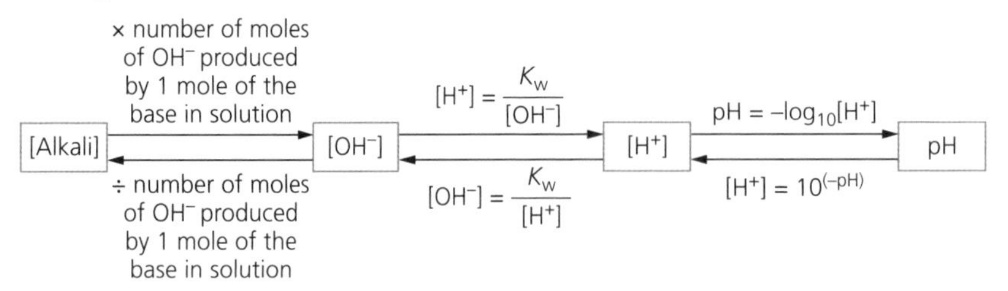

Figure 23 Determining pH of a strong alkali

Worked example 1

Calculate the pH of a 0.644 mol dm^{-3} solution of sodium hydroxide ($K_w = 1.00 \times 10^{-14}$ mol^2 dm^{-6}). Give your answer to 2 decimal places.

Answer

[NaOH] = 0.644 mol dm^{-3}

[OH$^-$] = 0.644 mol dm^{-3} (as 1 OH$^-$ per NaOH)

$$[H^+] = \frac{K_w}{[OH^-]} = \frac{1.00 \times 10^{-14}}{0.644} = 1.55 \times 10^{-14}\ \text{mol dm}^{-3}$$

$$pH = -\log_{10}[H^+] = -\log_{10}(1.55 \times 10^{-14}) = 13.81$$

Worked example 2

Calculate the concentration of potassium hydroxide solution, in mol dm^{-3}, at 40°C, which has a pH of 13.64. K_w at 40°C is 2.92×10^{-14} mol^2 dm^{-6}. Give your answer to 3 significant figures.

Answer

$$[H^+] = 10^{-13.64} = 2.29 \times 10^{-14}\ \text{mol dm}^{-3}$$

$$[OH^-] = \frac{K_w}{[H^+]} = \frac{2.92 \times 10^{-14}}{2.29 \times 10^{-14}} = 1.28\ \text{mol dm}^{-3}$$

[KOH] = [OH$^-$]

So:

[KOH] = 1.28 mol dm^{-3}

Exam tip

The majority of strong bases (alkalis) you will encounter will have 1 mole of OH$^-$ ions per mole of the base. However, sometimes a question is set on a group 2 hydroxide where you are asked to assume that the base is strong and so the number of moles of OH$^-$ per mole of base for $Ca(OH)_2$ is 2.

Check the second calculation the other way round if you have time. Make sure you get a pH of 13.64 using a concentration of 1.28 mol dm^{-3} and the K_w at that temperature. The answer may be approximate, but this may be to do with rounding.

pH of weak acids

Weak acids are slightly dissociated in solution. This is represented using a reversible arrow ($\rightleftharpoons$). The equilibrium constant for the acid dissociation is represented by K_a.

For the general weak acid dissociation equation, $HA \rightleftharpoons H^+ + A^-$.

$K_a = \frac{[H^+][A^-]}{[HA]}$ for the species present at equilibrium.

K_a always has units of $mol\,dm^{-3}$.

Worked example

Write an expression for K_a for ethanoic acid, CH_3COOH.

Answer

$$K_a = \frac{[H^+][CH_3COO^-]}{[CH_3COOH]}$$

Exam tip

HA is the undissociated acid. Remember that the undissociated acid does not cause it to be acidic. It is the concentration of H^+ that causes acidity when the acid dissociates. HA is not acidic until it dissociates. It is important to be able to write K_a expressions for weak acids. It is always the [anion] and [H^+] on the top and [undissociated acid] on the bottom.

Knowledge check 19

Write a K_a expression for methanoic acid, HCOOH.

The value of K_a gives a measure of the strength of the acid. A higher K_a value indicates a 'stronger' weak acid. For example, ethanoic acid has a K_a value of $1.8 \times 10^{-5}\,mol\,dm^{-3}$, whereas hydrocyanic acid has a K_a of $4.9 \times 10^{-10}\,mol\,dm^{-3}$. Ethanoic acid is a 'stronger' acid than hydrocyanic acid.

For a weak acid, [H^+] is calculated from the concentration of the acid and the K_a value using the expression:

$$[H^+] = \sqrt{K_a \times \text{initial concentration of acid}}$$

The pH is calculated using $pH = -\log_{10}[H^+]$. This is summarised in Figure 24.

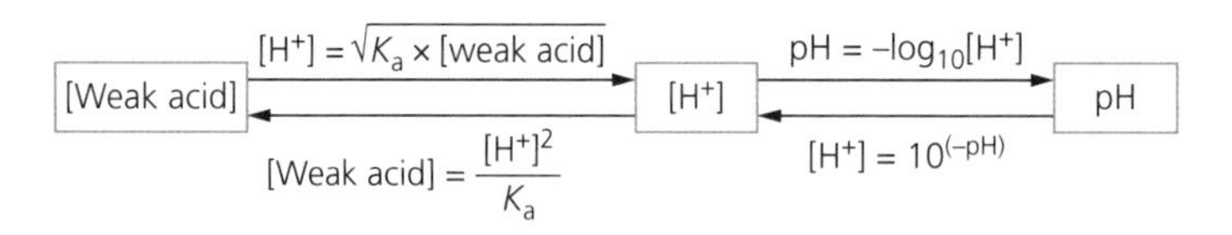

Figure 24 Determining pH of a weak acid

pK_a

The pK_a of a weak acid may be given in place of its K_a. $pK_a = -\log_{10}K_a$. To convert between a pK_a value and K_a use the following: $K_a = 10^{(-pK_a)}$.

The higher the pK_a value, the weaker the acid. A lower pK_a indicates a 'stronger' acid.

Worked example 1

K_a for methanoic acid (CH_3COOH) is 1.60×10^{-4} mol dm^{-3}. Calculate the pH of a 0.128 mol dm^{-3} solution of methanoic acid. Give your answer to 2 decimal places.

Answer

$$[H^+] = \sqrt{K_a \times [\text{weak acid}]}$$

$$[H^+] = \sqrt{1.60 \times 10^{-4} \times 0.128} = \sqrt{2.048 \times 10^{-5}} = 4.525 \times 10^{-3}\ \text{mol dm}^{-3}$$

$$\text{pH} = -\log_{10}[H^+] = -\log_{10}(4.525 \times 10^{-3}) = 2.34$$

Worked example 2

Calculate the pH of a solution if 0.0253 moles of a weak acid is dissolved in 500 cm^3 of water (pK_a for the acid = 2.85). Give your answer to 2 decimal places.

Answer

Concentration of the weak acid in mol dm^{-3} = 0.0253 × 2 = 0.0506 mol dm^{-3}.

$$K_a = 10^{(-pKa)} = 10^{(-2.85)} = 1.41 \times 10^{-3}\ \text{mol dm}^{-3}$$

$$[H^+] = \sqrt{K_a \times [\text{weak acid}]}$$

$$[H^+] = \sqrt{1.41 \times 10^{-3} \times 0.0506} = \sqrt{7.13 \times 10^{-5}} = 8.45 \times 10^{-3}\ \text{mol dm}^{-3}$$

$$\text{pH} = -\log_{10}[H^+] = -\log_{10}(8.45 \times 10^{-3}) = 2.07$$

The acid dissociation constant, K_a, may be calculated from pH and the concentration of the weak acid. $[H^+]$ is determined from pH in the usual way and the acid dissociation constant can be calculated using $K_a = [H^+]^2/[\text{weak acid}]$.

Worked example 1

A 0.0244 mol dm^{-3} solution of a weak acid has a pH of 3.24. Calculate a value for the acid dissociation constant, K_a, for this weak acid. Give your answer to 3 significant figures.

Answer

$$[H^+] = 10^{-3.24} = 5.75 \times 10^{-4}\ \text{mol dm}^{-3}$$

$$K_a = \frac{[H^+]^2}{[\text{weak acid}]} = \frac{(5.75 \times 10^{-4})^2}{0.0244} = 1.36 \times 10^{-5}\ \text{mol dm}^{-3}.$$

The concentration of the weak acid may be determined from the pH and the K_a using the expression $[\text{weak acid}] = [H^+]^2/K_a$.

Worked example 2

The pK_a for nitrous acid, HNO_2, is 3.25. A solution of nitrous acid has a pH of 2.35. Calculate the concentration of the solution of nitrous acid. Give your answer to 3 significant figures.

Answer

$[H^+] = 10^{-2.35} = 4.47 \times 10^{-3}\,\text{mol dm}^{-3}$

$K_a = 10^{-pK_a} = 10^{-3.25} = 5.62 \times 10^{-4}\,\text{mol dm}^{-3}$

$$[\text{weak acid}] = \frac{[H^+]^2}{K_a} = \frac{(4.47 \times 10^{-3})^2}{5.62 \times 10^{-4}} = 0.0356\,\text{mol dm}^{-3}$$

Knowledge check 20

Calculate the pH of a weak acid of concentration $0.0145\,\text{mol dm}^{-3}$, where $K_a = 1.44 \times 10^{-4}\,\text{mol dm}^{-3}$.

Dilutions

Diluting a solution of a strong acid or a strong base will change the pH as the concentration of H^+ or OH^- ions will change.

The main expression to remember is the one below as this allows the concentration of the diluted solution to be determined.

$$\text{concentration of diluted solution (mol dm}^{-3}) = \frac{\text{moles of solute}}{\text{volume of new solution (cm}^3)} \times 1000$$

Dilutions of strong acids

Worked example

$17.2\,\text{cm}^3$ of a $1.24\,\text{mol dm}^{-3}$ solution of nitric acid are dilute to $150\,\text{cm}^3$. Calculate the pH of the solution formed. Give your answer to 2 decimal places.

Answer

$$\text{Moles of } HNO_3 \text{ added} = \frac{17.2 \times 1.24}{1000} = 0.0213\,\text{mol}$$

New volume = $150\,\text{cm}^3$

$$\text{Concentration of new solution} = \frac{0.0213}{150} \times 1000 = 0.142\,\text{mol dm}^{-3}$$

Nitric acid is monoprotic. So $[H^+] = [HNO_3]$:

$[H^+] = 0.142\,\text{mol dm}^{-3}$

$pH = -\log_{10}[H^+] = -\log_{10}(0.142) = 0.85$

Exam tip

If the acid was diprotic like H_2SO_4:
$[H^+] = 2 \times [H_2SO_4]$

The volume of water added or the total volume of the solution may also be calculated for a given pH of a diluted solution.

Worked example

Calculate the volume of water added to $40.0\,cm^3$ of a $0.0278\,mol\,dm^{-3}$ solution of sulfuric acid so that the pH of the resulting solution is 1.88. Give your answer to 3 significant figures.

Answer

$$\text{Moles of } H_2SO_4 \text{ in original solution} = \frac{40.0 \times 0.0278}{1000} = 1.11 \times 10^{-3}\,mol$$

$$\text{Moles of } H^+ \text{ in original solution} = 2.22 \times 10^{-3}\,mol$$

$$\text{New } [H^+] = 10^{-1.88} = 0.0132\,mol\,dm^{-3}$$

$$\text{New } [H^+] = \frac{\text{moles of } H^+}{\text{volume of new solution } (cm^3)} \times 1000$$

$$\text{Volume of new solution} = \frac{\text{moles of } H^+}{\text{new } [H^+]} \times 1000$$

$$\text{Volume of new solution} = \frac{2.22 \times 10^{-3}}{0.0132} \times 1000 = 168\,cm^3$$

$$\text{Volume of water added} = 168 - 40.0 = 128\,cm^3$$

Dilution of solutions of strong bases

For a solution of a strong base, K_w is required to determine the pH.

Worked example

$14.8\,cm^3$ of $0.884\,mol\,dm^{-3}$ sodium hydroxide solution were diluted to $200\,cm^3$. Calculate the pH of the new solution at 25°C. Give your answer to 2 decimal places.

Answer

$$K_w = 1.00 \times 10^{-14}\,mol^2\,dm^{-6} \text{ at } 25°C.$$

$$\text{Moles of NaOH} = \frac{14.8 \times 0.884}{1000} = 0.0131\,mol$$

$$\text{New [NaOH]} = \frac{0.0131}{200} \times 1000 = 0.0655\,mol\,dm^{-3}$$

$$[NaOH] = [OH^-] = 0.0655\,mol\,dm^{-3}$$

$$[H^+] = \frac{K_w}{[OH^-]} = \frac{1.00 \times 10^{-14}}{0.0655} = 1.53 \times 10^{-13}\,mol\,dm^{-3}$$

$$pH = -\log_{10}[H^+] = -\log_{10}(1.53 \times 10^{-13}) = 12.82$$

Neutralisations

Adding acid to base or vice versa will mean some of the acid or base has been neutralised and this will also change the pH. In this type of question it is important to be able to calculate the amount, in moles, of the acid or base that is left over.

Worked example 1

19.2 cm^3 of 0.322 mol dm^{-3} hydrochloric acid were added to 15.0 cm^3 of 0.350 mol dm^{-3} solution of sodium hydroxide. Calculate the pH of the resulting solution. Give your answer to 2 decimal places.

Answer

$$NaOH + HCl \rightarrow NaCl + H_2O$$

NaOH and HCl react in a 1:1 ratio.

$$\text{Initial moles of NaOH} = \frac{15.0 \times 0.350}{1000} = 5.25 \times 10^{-3}\,\text{mol}$$

$$\text{Initial moles of HCl} = \frac{19.2 \times 0.322}{1000} = 6.18 \times 10^{-3}\,\text{mol}$$

HCl is in excess. Moles of HCl remaining $= 6.18 \times 10^{-3} - 5.25 \times 10^{-3} = 9.30 \times 10^{-4}$ mol

New total volume of solution = 19.2 + 15.0 = 34.2 cm^3

$$\text{New concentration of reactant in excess} = \frac{\text{moles of reactant in excess}}{\text{volume of new solution (cm}^3\text{)}} \times 1000$$

$$\text{New concentration of reactant (HCl) in excess} = \frac{9.30 \times 10^{-4}}{34.2} \times 1000$$

$$= 0.0272\,\text{mol dm}^{-3}$$

HCl is monoprotic. So $[H^+] = [HCl]$:

$$[H^+] = 0.0272\,\text{mol dm}^{-3}$$

$$pH = -\log_{10}[H^+] = -\log_{10}(0.0272) = 1.57$$

Be careful with 1:2 or 2:1 ratios of acid:base.

Worked example 2

22.5 cm^3 of 0.0228 mol dm^{-3} sulfuric acid is reacted with 27.4 cm^3 of 0.0544 mol dm^{-3} potassium hydroxide solution. Calculate the pH of the solution formed at 25°C. Give your answer to 2 decimal places. $K_w = 1.00 \times 10^{-14}\,\text{mol}^2\,\text{dm}^{-6}$ at 25°C.

Answer

$$\text{Initial moles of KOH} = \frac{27.4 \times 0.0544}{1000} = 1.49 \times 10^{-3}\,\text{mol}$$

$$\text{Initial moles of } H_2SO_4 = \frac{22.5 \times 0.0228}{1000} = 5.13 \times 10^{-4}\,\text{mol}$$

$$2KOH + H_2SO_4 \rightarrow K_2SO_4 + 2H_2O$$

→

KOH reacts with H_2SO_4 in a 2:1 ratio.

$$2 \times 5.13 \times 10^{-4} = 1.03 \times 10^{-3}\,\text{mol}$$

so KOH is in excess.

$$\text{Moles of KOH in excess} = 1.49 \times 10^{-3} - 1.03 \times 10^{-3} = 4.60 \times 10^{-4}$$

$$\text{Total volume} = 22.5 + 27.4 = 49.9\,\text{cm}^3$$

$$\text{New concentration of KOH} = \frac{\text{moles of reactant in excess}}{\text{volume of new solution (cm}^3)} \times 1000$$

$$\text{New [KOH]} = \frac{4.60 \times 10^{-4}}{49.9} = 9.22 \times 10^{-3}\,\text{mol dm}^{-3}$$

As KOH has one OH^-, $[OH^-] = [KOH] = 9.22 \times 10^{-3}$:

$$[H^+] = \frac{K_w}{[OH^-]} = \frac{1.00 \times 10^{-14}}{9.22 \times 10^{-3}} = 1.08 \times 10^{-12}\,\text{mol dm}^{-3}$$

$$\text{pH} = -\log_{10}[H^+] = -\log_{10}(1.08 \times 10^{-12}) = 11.97$$

Exam tip

All values in these calculations were stated to 3 significant figures as that is the level of accuracy stated throughout. However, even doing all the calculations on a calculator gives the same pH.

Knowledge check 21

Calculate the pH of the solution obtained from diluting 10.0 cm^3 of a 0.145 mol dm^{-3} solution of sulfuric acid to 500.0 cm^3.

pH curves

A pH curve is a graph of pH against volume of alkali or acid added. It is obtained using a pH meter or probe during a titration.

- A typical curve shows the initial pH of the acid or alkali in the conical flask and the vertical region of the curve occurs at a volume where neutralisation occurs.
- The shape of the curve shows the type of titration. The vertical region occurs when one drop is added from the burette, which neutralises the substance in the conical flask.
- The pH changes rapidly and any indicator chosen for the titration must have a range of pH, which changes completely in the pH range of this vertical region so that the full colour change of the indicator is shown.

Four types of sample titration curves are shown in Figure 25 with some notes below each one as to the starting pH and the shape of the curve including the pH range of the vertical region.

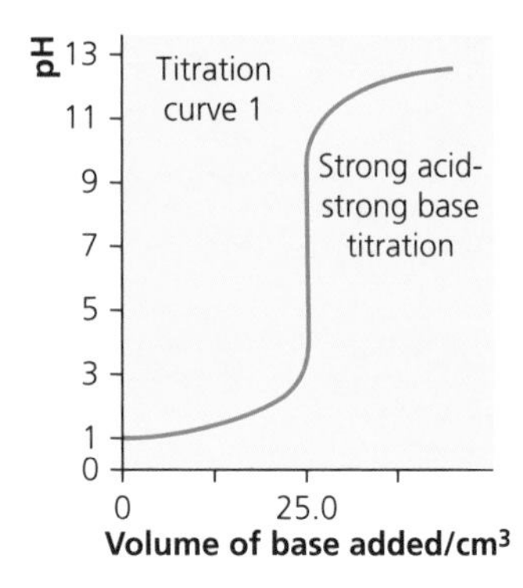

- Vertical region from pH 3 to pH 10.
- Initial pH is 1, so a strong acid in the conical flask at the start.
- 25.0 cm^3 of the base solution are required to neutralise the acid.
- This is the curve for a **strong acid–strong base titration**.

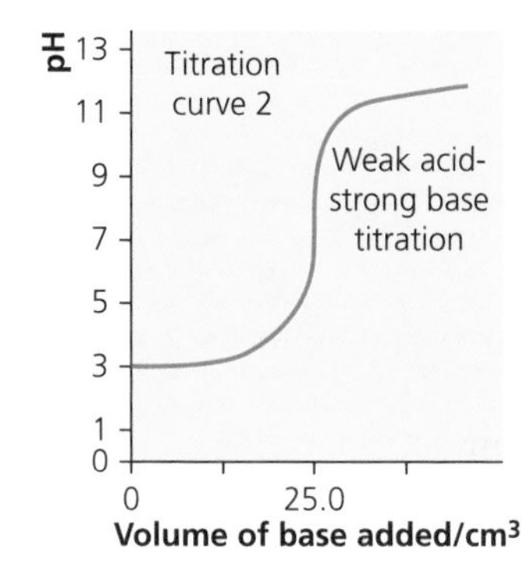

- Vertical region from pH 6 to pH 10.
- Initial pH is 3, so a weak acid in the conical flask at the start.
- 25.0 cm^3 of the base solution are required to neutralise the acid.
- This is the curve for a **weak acid–strong base titration**.

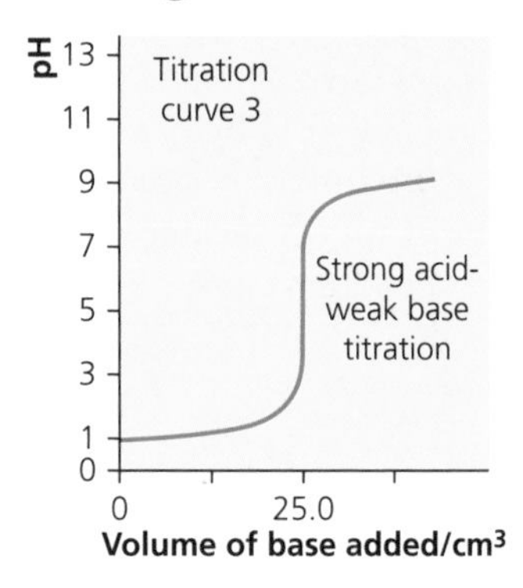

- Vertical region from pH 3 to pH 8.
- Initial pH is 1, so a strong acid in the conical flask at the start.
- 25.0 cm^3 of the base solution are required to neutralise the acid.
- This is the curve for a **strong acid–weak base titration**.

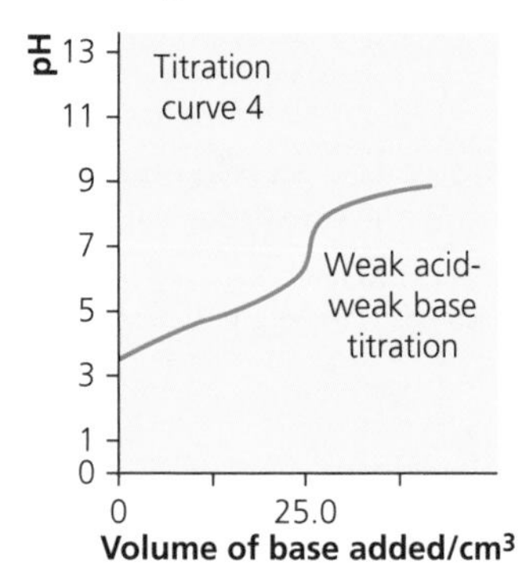

- No clear vertical region.
- Initial pH is greater than 3, so a weak acid in the conical flask at the start.
- It is not clear what volume of the base solution is required to neutralise the acid.
- This is the curve for a **weak acid–weak base titration**.

Figure 25 Titration curves

Sketching titration curves

If you are asked to sketch a titration curve you may be given:

- An initial calculation for the pH of the strong or weak acid — this is the starting pH for the curve where the volume of alkali added is equal to 0.00 cm^3.
- The names of the acid and base used — this will determine the shape of the titration curve including the length of the point of inflection.

Exam tip

The information on the shapes of the curves is important, as the shape of the curve depends on the type (strong or weak) of acid or base (alkali) used in the titration. It would be very unusual for it to be a weak acid–weak base (alkali) titration. The sketch is just that — a rough sketch of the shape — so follow the (rough) guidelines as to where the inflection point (vertical section) occurs, i.e. between what approximate pH values.

- The concentrations of the acid and base (alkali) used — the volume at which the point of inflection occurs can be calculated from the volume of base (alkali) required to neutralise the acid.

Worked example

In a titration 0.175 mol dm^{-3} sodium hydroxide solution is added to 25.0 cm^3 of 0.123 mol dm^{-3} hydrochloric acid. Sketch the titration curve you would expect to obtain.

Answer

Initial pH:

HCl is a monobasic acid. So $[H^+] = 0.123$ mol dm^{-3}.

Initial pH $= -\log_{10}[H^+] = -\log_{10}(0.123) = 0.91$.

Shape:

Strong acid–strong base titration. So curve shaped like curve 1 — equivalence point with a change between approximately pH 3 and 10.

Volume of base:

$$\text{Moles of HCl present} = \frac{25.0 \times 0.123}{1000} = 3.08 \times 10^{-3}\,\text{mol}$$

$$NaOH + HCl \rightarrow NaCl + H_2O$$

1 mole of NaOH reacts with 1 mole of HCl.

$$\text{Moles of NaOH} = 3.08 \times 10^{-3}\,\text{mol}$$

$$\text{Volume of NaOH} = \frac{3.08 \times 10^{-3} \times 1000}{0.175} = 17.6\,\text{cm}^3$$

The titration curve for this neutralisation looks like Figure 26.

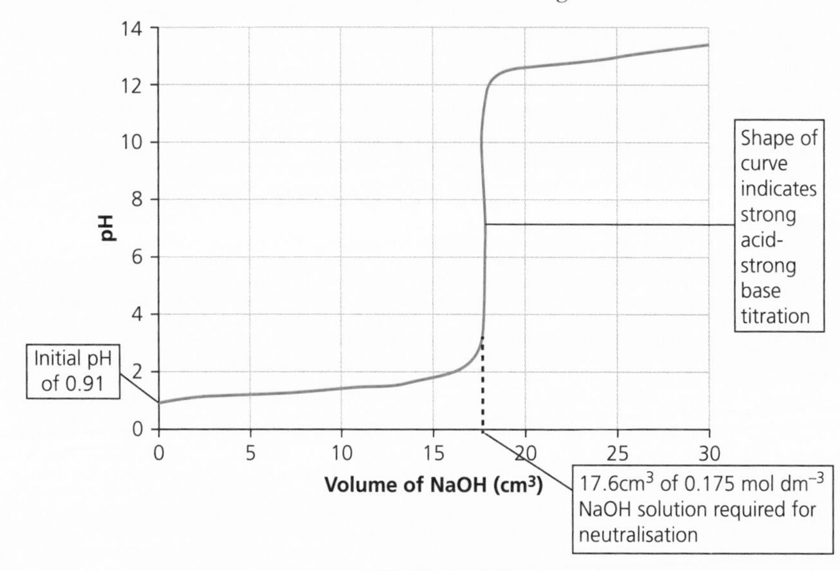

Figure 26

Exam tip

The titration curve may be for a volume of acid added to a volume of alkali. Work out the initial pH of the strong alkali — this is the initial pH; the volume required for neutralisation is determined in the same way; the shape of curve is the same but reversed.

Indicators for titrations

The colour change pH range of the indicator must be completely within the vertical region of the titration curve (Table 6).

Table 6 Some indicators and their colour change pH ranges

Indicator	pH range of colour change	Suitable for titrations between
Methyl orange	3.1 – 4.4	strong acid–strong base strong acid–weak base
Methyl red	4.4 – 6.2	strong acid–strong base strong acid–weak base
Phenolphthalein	8.3 – 10.0	strong acid–strong base weak acid–strong base

Table 7 Colours of common indicators

Indicator	Colour change when adding acid to alkali	Colour when adding alkali to acid
Methyl orange	Yellow to red	Red to yellow
Phenolphthalein	Pink to colourless	Colourless to pink

Required practical 9

Investigating pH changes

Students should monitor pH using a pH meter or a pH probe in the reaction between a weak acid and a strong base and between a strong acid and a weak base.

Buffers

Buffers are solutions that maintain almost constant pH despite dilution or the addition of small amounts of acid or base.

- An acidic buffer is a solution formed from a weak acid and its salt.
- A typical acidic buffer would be a solution containing ethanoic acid and the ethanoate ion.
- A basic buffer is a solution of a weak base and its salt.
- A typical basic buffer would be a solution containing ammonia and the ammonium ion.

Dilution of a buffer changes the concentration of the ions to the same degree and this has little effect on the pH of the buffer.

Action of buffers

Buffers maintain an almost constant pH by being able to remove small quantities of H^+ ions or OH^- ions added to them.

Exam tip

Methyl orange and phenolphthalein are the main indicators of choice. Both may be used for strong acid–strong base titrations. Methyl orange is used for strong acid–weak base (alkali) titrations. Phenolphthalein is used for weak acid–strong base (alkali) titrations.

Exam tip

A pH meter must be used for weak acid–weak base (alkali) titrations as there is no vertical region to the pH curve.

Knowledge check 22

State the colour of phenolphthalein in a solution of pH 12.00.

Acidic buffers

The anion (A^-) and the undissociated weak acid can react with added H^+ and OH^- ions respectively to maintain an almost constant pH.

Explanation of buffering action

Addition of acid: extra H^+ added:

A^- reacts with H^+ to form HA ($A^- + H^+ \rightarrow HA$).

This removes the added H^+.

Addition of base: extra OH^- ions added:

HA reacts with OH^-, forming A^- and H_2O:

($HA + H_2O \rightarrow A^- + H_2O$)

This removes the added OH^-.

Basic buffers

The weak base Y and its ion YH^+ can react with added H^+ and OH^- ions respectively to maintain an almost constant pH.

Addition of acid: extra H^+ ions added:

Y reacts with H^+ to form YH^+:

($Y + H^+ \rightarrow YH^+$)

This removes the added H^+.

Addition of base: extra OH^- ions added:

YH^+ reacts with OH^- to form Y and H_2O:

($YH^+ + OH^- \rightarrow Y + H_2O$)

This removes the added OH^-.

Exam tip

Again use the given base. For example, if the basic buffer is made from ammonia (NH_3) and ammonium ions (NH_4^+), use these in place of Y and YH^+ in the equations above. Amines are weak bases and can also be used to form a basic buffer, for example, methylamine (CH_3NH_2) and the methylammonium ion ($CH_3NH_3^+$).

Exam tip

If the weak acid and its salt are named, then use the correct anion and weak acid. For example, if the buffer is made from ethanoic acid and ethanoate ions, the equation should read $CH_3COO^- + H^+ \rightarrow CH_3COOH$ to show the removal of the added H^+ ions.

Exam tip

Again if the weak acid and its salt are named, then use the correct weak acid and anion. For example, if the buffer is made from hydrocyanic acid and cyanide ions, the equation should read $HCN + OH^- \rightarrow CN^- + H_2O$ to show the removal of the added OH^- ions.

Preparation of an acidic buffer

There are several ways to prepare an acidic buffer, but all of them produce a solution containing a weak acid and its anion, for example, ethanoic acid and the ethanoate ion.

Methods include:

- adding solid salt of a weak acid to a solution of the weak acid
- mixing a solution of a weak acid with a solution of its salt
- mixing a strong base solution with an excess of a solution of a weak acid

Calculating pH of a buffer

To calculate the pH of a buffer you must determine the concentration of the anion $[A^-]$ (this equals the concentration of the salt) and the concentration of the weak acid [HA] from the information given in the method used to prepare the buffer.

$[H^+]$ is calculated from the K_a expression and pH is then calculated in the usual way.

Worked example 1

0.0185 mol of sodium ethanoate were dissolved in $150\,cm^3$ of $0.0242\,mol\,dm^{-3}$ ethanoic acid ($K_a = 1.74 \times 10^{-5}\,mol\,dm^{-3}$). Calculate the pH of this buffer solution. Give your answer to 2 decimal places.

Answer

Concentration of weak acid $[CH_3COOH] = 0.0242\,mol\,dm^{-3}$

Exam tip

Remember to always try to calculate the concentration of the weak acid and its salt. The calculation always follows a similar pattern once you have calculated these concentrations. In this example the concentration of the weak acid is unchanged as solid is added. So there is no dilution of the weak acid solution.

Concentration of salt $[CH_3COONa] = \frac{0.0185}{150} \times 1000 = 0.123\,mol\,dm^{-3}$

Concentration of anion $[CH_3COO^-] = 0.123\,mol\,dm^{-3}$

$$K_a = \frac{[CH_3COO^-][H^+]}{[CH_3COOH]} = 1.74 \times 10^{-5}\,mol\,dm^{-3}$$

$$[H^+] = \frac{K_a \times [CH_3COOH]}{[CH_3COO^-]} = \frac{1.74 \times 10^{-5} \times 0.0242}{0.123} = 3.42 \times 10^{-6}\,mol\,dm^{-3}$$

$$pH = -\log_{10}[H^+] = -\log_{10}(3.42 \times 10^{-6}) = 5.47$$

Exam tip

The concentration of the anion in $mol\,dm^{-3}$ is determined by dividing the number of moles of salt (the solute) by the volume of the solution and multiplying by 1000 to convert to the number of moles per $1\,dm^3$ ($1000\,cm^3$).

Exam tip

All methods of determining the pH of a buffer end in the same way — using the concentration of the acid [HA] and the concentration of the anion $[A^-]$ to determine $[H^+]$ from K_a. pH is calculated from $[H^+]$.

Worked example 2

14.2 cm^3 of a 0.224 mol dm^{-3} solution of the salt of a weak acid, NaX, is added to 50.0 cm^3 of a 0.482 mol dm^{-3} solution of the weak acid, HX. The pK_a for HX is 4.23. Calculate the pH of the buffer formed. Give your answer to 2 decimal places.

Answer

$$\text{Moles of NaX added} = \frac{14.2 \times 0.224}{1000} = 3.18 \times 10^{-3}\,\text{mol}$$

$$[\text{NaX}] = [\text{X}^-] = \frac{3.18 \times 10^{-3}}{64.2} \times 1000 = 0.0495\,\text{mol dm}^{-3}$$

$$\text{Moles of HX added} = \frac{50.0 \times 0.482}{1000} = 0.0241\,\text{mol}$$

$$[\text{HX}] = \frac{0.0241}{64.2} \times 1000 = 0.375\,\text{mol dm}^{-3}$$

$$K_a = 10^{-\text{p}K_a} = 10^{-4.23} = 5.89 \times 10^{-5}\,\text{mol dm}^{-3}$$

$$K_a = \frac{[\text{X}^-][\text{H}^+]}{[\text{HX}]} = 5.89 \times 10^{-5}\,\text{mol dm}^{-3}$$

$$[\text{H}^+] = \frac{K_a \times [\text{HX}]}{[\text{X}^-]} = \frac{5.89 \times 10^{-5} \times 0.375}{0.0495} = 4.47 \times 10^{-4}\,\text{mol dm}^{-3}$$

$$\text{pH} = -\log_{10}[\text{H}^+] = -\log_{10}(4.47 \times 10^{-4}) = 3.35$$

Exam tip

The pK_a is converted into a K_a using $K_a = 10^{(-\text{p}K_a)}$. The new concentrations in mol dm^{-3} are determined using the number of moles of solute in the new total volume of 64.2 cm^3 (50.0 + 14.2) converted to 1 dm^3 (1000 cm^3).

Worked example 3

45.0 cm^3 of 0.0500 mol dm^{-3} methanoic acid were mixed with 55.0 cm^3 of 0.0270 mol dm^{-3} sodium hydroxide solution. The K_a for methanoic acid is 1.60 × 10^{-4} mol dm^{-3}. Calculate the pH of the resulting buffer solution. Give your answer to 2 decimal places.

Answer

$$\text{Moles of methanoic acid added (HCOOH)} = \frac{45.0 \times 0.0500}{1000} = 2.25 \times 10^{-3}\,\text{mol}$$

$$\text{Moles of sodium hydroxide added (NaOH)} = \frac{55.0 \times 0.0270}{1000} = 1.49 \times 10^{-3}\,\text{mol}$$

The equation for the reaction between methanoic acid and sodium hydroxide solution is:

$$\text{HCOOH} + \text{NaOH} \rightarrow \text{HCOONa} + \text{H}_2\text{O}$$

$$\text{Moles of HCOOH after reaction} = 2.25 \times 10^{-3} - 1.49 \times 10^{-3} = 7.60 \times 10^{-4}\,\text{mol}$$

$$\text{Moles of HCOONa after reaction} = 1.49 \times 10^{-3}\,\text{mol}$$

$$[\text{HCOOH}] = \frac{7.60 \times 10^{-3}}{100} \times 1000 = 0.0760\,\text{mol dm}^{-3}$$

Exam tip

Again both solutions dilute each other and the sodium hydroxide reacts with some of the methanoic acid to form the salt, sodium methanoate (HCOONa). The new total volume in this example is 100 cm^3 and this will be used to calculate the new concentrations of the weak acid and its salt.

$$[\text{HCOONa}] = [\text{HCOO}^-] = \frac{1.49 \times 10^{-3}}{100} \times 1000 = 0.0149\,\text{mol dm}^{-3}$$

$$K_a = \frac{[\text{HCOO}^-][\text{H}^+]}{[\text{HCOOH}]} = 1.60 \times 10^{-4}\,\text{mol dm}^{-3}$$

$$[\text{H}^+] = \frac{K_a \times [\text{HCOOH}]}{[\text{HCOO}^-]} = \frac{1.60 \times 10^{-4} \times 0.0760}{0.0149} = 8.16 \times 10^{-4}\,\text{mol dm}^{-3}$$

$$\text{pH} = -\log_{10}[\text{H}^+] = -\log_{10}(8.16 \times 10^{-4}) = 3.09$$

Exam tip

All of the sodium hydroxide has been used up, leaving only some moles of the weak acid (HCOOH) and some moles of the salt (HCOONa) in solution. The water formed is in the solution.

Henderson–Hasselbalch equation

The Henderson–Hasselbalch equation may also be used to calculate the pH of a buffer. $\text{pH} = \text{p}K_a + \log_{10}([\text{A}^-]/[\text{HA}])$, where $[\text{A}^-]$ is the concentration of the anion and [HA] is the concentration of the acid.

The example above is calculated as:

$$\text{p}K_a = -\log_{10}K_a = -\log_{10}(1.60 \times 10^{-4}) = 3.80$$

$$[\text{A}^-] = [\text{HCOO}^-] = 0.0149\,\text{mol dm}^{-3}$$

$$[\text{HA}] = [\text{HCOOH}] = 0.0760\,\text{mol dm}^{-3}$$

$$\text{pH} = 3.80 + \log_{10}\left(\frac{0.0149}{0.0760}\right) = 3.80 - 0.71 = 3.09$$

Exam tip

In some cases the concentration of the weak acid and its salt may be the same. In this case $[\text{H}^+] = K_a$.

Knowledge check 23

What is a buffer?

Addition of acid or alkali to a buffer

A buffer maintains an approximately constant pH when small amounts of acid or base are added to it. The new pH of the buffer may be calculated by calculating the change in the concentration of the anion and the weak acid.

When acid is added to an acidic buffer, the anion reacts with the H^+ ions that have been added to remove them:

$$\text{A}^- + \text{H}^+ \rightarrow \text{HA}$$

This means that the amount of A^- decreases by the same amount as the moles of H^+ added and the amount of HA increases by the same amount. These new amounts can be used to recalculate the pH of the buffer.

Worked example

a A buffer was prepared by mixing 0.0258 moles of ethanoic acid with 0.0370 moles of sodium ethanoate in 200 cm^3. K_a for ethanoic acid is $1.74 \times 10^{-5}\,\text{mol dm}^{-3}$. Calculate the pH of the buffer. Give your answer to 2 decimal places.

b 0.00172 moles of hydrochloric acid were added to the buffer in **a**. Calculate the new pH of the buffer. Give your answer to 2 decimal places.

→

Answer

a $[CH_3COOH] = \frac{0.0258}{200} \times 1000 = 0.129\,mol\,dm^{-3}$

$[CH_3COO^-] = \frac{0.0370}{200} \times 1000 = 0.185\,mol\,dm^{-3}$

$K_a = \frac{[CHCOO^-][H^+]}{[CHCOOH]} = 1.74 \times 10^{-5}\,mol\,dm^{-3}$

$[H^+] = \frac{K_a \times [CHCOOH]}{[CHCOO^-]} = \frac{1.74 \times 10^{-5} \times 0.129}{0.185} = 1.21 \times 10^{-5}\,mol\,dm^{-3}$

$pH = -\log_{10}[H^+] = -\log_{10}(1.21 \times 10^{-5}) = 4.92$

b HCl is monoprotic, so 0.00172 mol of H^+ added.

CH_3COO^- decreases by 0.00172, so new $CH_3COO^- = 0.0370 - 0.00172 = 0.0353$ mol

CH_3COOH increases by 0.00172, so new $CH_3COOH = 0.0258 + 0.00172 = 0.0275$ mol

$[CH_3COOH] = \frac{0.0275}{200} \times 1000 = 0.138\,mol\,dm^{-3}$

$[CH_3COO^-] = \frac{0.0353}{200} \times 1000 = 0.177\,mol\,dm^{-3}$

$[H^+] = \frac{K_a \times [CHCOOH]}{[CHCOO^-]} = \frac{1.74 \times 10^{-5} \times 0.138}{0.177} = 1.36 \times 10^{-5}\,mol\,dm^{-3}$

$pH = -\log_{10}[H^+] = -\log_{10}(1.36 \times 10^{-5}) = 4.87$

So the pH only changes from 4.92 to 4.87 on the addition of the hydrochloric acid.

Exam tip

The calculation is virtually the same for the addition of alkali except the equation $HA + OH^- \rightarrow A^- + H_2O$ is used. The amount of HA decreases by the same amount as the amount of OH^- ions added and the amount of A^- increases by the same amount. The rest of the calculation is identical.

Summary

- An acid is a proton donor; a base is a proton acceptor; a proton is a hydrogen ion (H^+).
- $pH = -\log_{10}[H^+]$, where $[H^+]$ represents the concentration of H^+ in $mol\,dm^{-3}$.
- The pH of a strong acid can be calculated from $[H^+]$ = concentration of the acid × proticity of the acid. Then $pH = -\log_{10}[H^+]$.
- $K_w = [H^+][OH^-] = 1.00 \times 10^{-14}\,mol^{-2}\,dm^6$ at 25°C.
- The pH of strong bases is calculated using K_w. $H^+ = K_w/[OH^-]$, where $[OH^-]$ is equal to the concentration of the base × number of moles of OH^- produced by 1 mole of the base in solution.
- For a weak acid $[H^+] = \sqrt{K_a \times [\text{weak acid}]}$.
- The indicator in a titration depends on the pH range of the vertical region in the pH curve.
- A buffer is a solution which maintains almost constant pH when small amounts of acid or alkali are added.
- An acidic buffer is a solution containing a weak acid and its salt; a basic buffer is a solution containing a weak base and its salt.
- The pH of an acidic buffer is calculated by determining the concentration of the weak acid and its anion in the solution and using the K_a expression to calculate $[H^+]$.

Questions & Answers

This section contains a mix of multiple-choice and structured questions similar to those you can expect to find in the A-level papers.

The examinations

The A-level examination consists of three examinations of 2 hours each. Papers 1 and 2 comprise short and long structured questions with a total mark of 105 each. Paper 3 covers the entire specification and contains practical and data analysis questions as well as multiple-choice questions worth 30 marks. For each multiple-choice question there is one correct answer and at least one clear distractor.

A-level paper 1 contains topics of physical chemistry (except 3.1.5 and 3.1.9) and includes inorganic chemistry (3.2). A-level paper 2 covers organic chemistry (3.3) and all the physical chemistry except 3.1.1, 3.1.7 and 3.1.8). A-level paper 3 covers the entire specification. The questions in this section are on physical chemistry sections 3.8 to 3.12 but also draw on sections 3.1 to 3.7 covered in the first student guide of this series.

About this section

Answers to questions are followed by comments, preceded by the icon e. Try the questions first to see how you get on and then check the answers and comments.

General tips

- Be accurate with your learning at this level — examiners will penalise incorrect wording.
- At least 20% of the marks in assessments for chemistry will require the use of mathematical skills. For any calculation, always follow it through to the end even if you feel you have made a mistake — there are marks for the correct method even if the final answer is incorrect.
- Always attempt to answer a multiple-choice question even if it is a guess (you have a 25% chance of getting it right).

The uniform mark you receive for each of paper 1 and paper 2 will be out of 105. The uniform mark for paper 3 is out of 90. The total marks for A-level Chemistry are 300.

Thermodynamics

Question 1

Which one of the following is exothermic?

A Enthalpy of atomisation

B First electron affinity

C First ionisation enthalpy

D Enthalpy of lattice dissociation (1 mark)

The answer is B ✓

ⓔ This type of question tests your knowledge of Born–Haber cycles and the associated enthalpy changes. Remember that the only enthalpy changes that are exothermic in a normal cycle are generally the enthalpy change of formation and the first electron affinity. Some enthalpy changes of formation are endothermic but this is unusual for ionic compounds as most are exothermic.

Question 2

Which one of the following reactions will show a decrease in entropy?

A $CaCO_3(s) \rightarrow CaO(s) + CO_2(g)$

B $CaCO_3(s) + 2HCl(aq) \rightarrow CaCl_2(aq) + CO_2(g) + H_2O(l)$

C $NH_3(g) + HCl(g) \rightarrow NH_4Cl(s)$

D $4NH_3(g) + 5O_2(g) \rightarrow 4NO(g) + 6H_2O(g)$ (1 mark)

The answer is C ✓

ⓔ Remember that entropy is a measure of disorder, so solids have lower enthalpy than liquids (and solutions), which have lower entropy than gases. Any reaction such as those in A and B that produce a gas from solids and solutions show an increase in entropy. Totally gaseous mixtures where the number of moles of gas increases also show an increase in entropy. Two gases mixing to form a solid as in C show a decrease in entropy. Remember that an increase in a single state — for example moles of a substance in solution increasing — causes an increase in entropy.

Question 3

Some enthalpy changes are given in Table 1.

Table 1

Equation	$\Delta H^{\ominus}$ (kJ mol^{-1})
$Cd(s) \rightarrow Cd(g)$	+113
$Cd(g) \rightarrow Cd^{+}(g) + e^{-}$	+870
$Cd^{+}(g) \rightarrow Cd^{2+}(g) + e^{-}$	+1600
$Cl_2(g) \rightarrow 2Cl(g)$	+242
$Cl(g) + e^{-} \rightarrow Cl^{-}(g)$	−348
$Cd(s) + Cl_2(g) \rightarrow CdCl_2(s)$	−392

(a) Use the data in Table 1 to calculate a value for the lattice dissociation enthalpy for cadmium chloride. (3 marks)

$\Delta_L H$ = +392 + 113 + 870 + 1600 + 242 + 2(−348) ✓✓

= + 2521 kJ mol^{-1} ✓

ⓔ The most common error with this type of 2+ metal ion halide is to forget to multiply the electron affinity of chlorine by 2. This would give an answer of +2869 kJ mol^{-1}, which would be worth 2 marks instead of 3.

(b) Table 2 gives the experimental and theoretical lattice enthalpies for cadmium bromide and potassium bromide. The theoretical values are calculated using the perfect ionic model.

Table 2

	Cadmium bromide	Potassium bromide
$\Delta_L H$ (theoretical)	+2481	+679
$\Delta_L H$ (experimental)	+2725	+678

(i) Explain what is meant by the perfect ionic model. (1 mark)

Ions are point charges/ions are perfect spheres/only electrostatic interactions between the ions/only ionic bonding with no covalent bonding/no polarisation of ions ✓

ⓔ Any of the above points can be used to explain the use of the perfect ionic model to calculate the theoretical value for the lattice enthalpy. This assumes that the ions are simply a perfect sphere with a point charge, so there are only ionic attractions between them. Polarisation of the ions can lead to some covalent character in the compound.

(ii) **Using the values in Table 2, comment on the bonding in each of the compounds.** (4 marks)

The bonding in potassium bromide is ionic. ✓

There is some covalent character to cadmium bromide. ✓

The values are close together for potassium bromide. ✓

The higher experimental value for cadmium bromide indicates some covalent bonding. ✓

ⓔ A common question is to compare the experimental lattice enthalpy with the theoretical one. The experimental lattice enthalpy is determined using the values in a Born–Haber cycle. An experimental value that is higher than the theoretical value suggests some additional bonding in the compound, which would suggest some covalent character. If the theoretical value is similar to the experimental value, then the compound is completely ionic.

Question 4

Silver(I) nitrate, $AgNO_3$, decomposes on heating according to the equation

$$2AgNO_3(s) \rightarrow 2Ag(s) + 2NO_2(g) + O_2(g)$$

The standard enthalpies of formation and standard entropy values of the reactants and products in this reaction are given in Table 3.

Table 3

Substance	$\Delta_f H^\ominus$ (kJ mol^{-1})	$S^\ominus$ (J K^{-1} mol^{-1})
$AgNO_3(s)$	−123	141
$Ag(s)$	0	43
$NO_2(g)$	+33	To be calculated
$O_2(g)$	0	205

(a) **Calculate the standard enthalpy change of reaction.** (2 marks)

(b) **Explain why the standard enthalpies of formation of $Ag(s)$ and $O_2(g)$ are zero.** (1 mark)

(c) **Calculate the standard entropy of $NO_2(g)$ if the $\Delta S^\ominus$ for the reaction is +489 J K^{-1} mol^{-1}.** (2 marks)

(d) **Determine the temperature, in kelvin, at which this decomposition becomes feasible.** (3 marks)

ⓔ This is a common type of synoptic question from AS and you need to revise this from the energetics section. The calculation in (a) can be done using Hess's law.

(a) Using enthalpies of formation the enthalpy change for the reaction ΔH is calculated using:

$\Delta H = \Sigma\Delta_f H^\ominus_{products} - \Sigma\Delta_f H^\ominus_{reactants}$

$\Delta H = 2(+33) - 2(-123)$ ✓ $= +312\,kJ\,mol^{-1}$ ✓

(b) Ag(s) and $O_2(g)$ are elements in their standard states. ✓

(c) $\Delta S^\ominus = \Sigma S^\ominus_{products} - \Sigma S^\ominus_{reactants}$

$+489 = 2(43) + 2(x) + 205 - 2(141)$ ✓

$2x = 480 \Rightarrow x = 240$ ✓

(d) For the reaction to be feasible $\Delta G^\ominus \leq 0$, so using $\Delta G^\ominus = 0$:

$\Delta G^\ominus = \Delta H^\ominus - T\Delta S^\ominus$

$\Delta H^\ominus = T\Delta S^\ominus$

$T = \frac{\Delta H^\ominus}{\Delta S^\ominus}$ ✓ $\frac{312}{0.489}$ ✓ $= 638\,K$ ✓

ⓔ In (d), remember to change the standard entropy change from $J\,K^{-1}\,mol^{-1}$ to $kJ\,K^{-1}\,mol^{-1}$ by dividing by 1000 because all the other units in the expression are in $kJ\,mol^{-1}$.

Question 5

The vaporisation of methanol is represented by the equation:

$$CH_3OH(l) \rightleftharpoons CH_3OH(g) \qquad \Delta H^\ominus = +35.3\,kJ\,mol^{-1}$$

The standard entropy of $CH_3OH(l)$ is $127\,J\,K^{-1}\,mol^{-1}$. Given that the reaction becomes feasible at 338 K, calculate the standard entropy of $CH_3OH(g)$. (6 marks)

$\Delta G^\ominus = 0$, so $\Delta H^\ominus = T\Delta S^\ominus$ ✓

$+35.3 = 338 \times \Delta S^\ominus$ ✓

$\Delta S^\ominus = \frac{35.3}{338} = +0.104\,kJ\,K^{-1}\,mol^{-1}$ ✓

$\Delta S^\ominus = +104\,J\,K^{-1}\,mol^{-1}$ ✓

$\Delta S^\ominus = S^\ominus_{products} - S^\ominus_{reactants}$

$+104 = S^\ominus_{products} - 127$ ✓

Entropy of $CH_3OH(g) = 104 + 127 = 231\,J\,K^{-1}\,mol^{-1}$ ✓

ⓔ For a system in equilibrium, $\Delta G = 0$. A change of state is considered a system in equilibrium.

Question 6

A graph of $\Delta G^\ominus$ against T is shown in Figure 1.

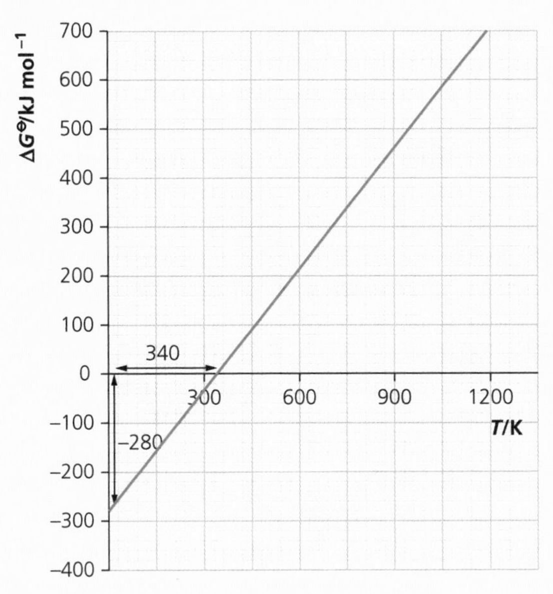

Figure 1

(a) From the graph state the value of $\Delta H^\ominus$ for this reaction. State the units. (1 mark)

$\Delta H^\ominus = -280\,\text{kJ}\,\text{mol}^{-1}$ ✓

ⓔ The value of $\Delta H^\ominus$ is equal to the value of $\Delta G^\ominus$ where the line hits the axis.

(b) From the graph calculate the value for $\Delta S^\ominus$ for this reaction. State the units. (3 marks)

gradient $= -\Delta H^\ominus = \frac{280}{340} = 0.824\,\text{kJ}\,\text{K}^{-1}\,\text{mol}^{-1}$ ✓

$\Delta S^\ominus = -824$ ✓ $\text{J}\,\text{K}^{-1}\,\text{mol}^{-1}$ ✓

ⓔ From the graph the gradient is $-\Delta S^\ominus$. The gradient of this graph is positive and the value is in $\text{kJ}\,\text{K}^{-1}\,\text{mol}^{-1}$. The sign must be changed to calculate $\Delta S^\ominus$ and also multiplied by 1000 to convert to $\text{J}\,\text{K}^{-1}\,\text{mol}^{-1}$.

(c) From the graph state the temperature below which the reaction is feasible. (1 mark)

$T = 340\,\text{K}$ ✓

ⓔ The point on the T axis where the line crosses is the temperature where a reaction changes from being feasible to not feasible — in this example as $\Delta G^\ominus$ becomes positive.

Rate equations

Question 1

The rate equation for the reaction A + 2B $\rightleftharpoons$ 3C + D

rate = k[A]^2

What are the units of the rate constant k?

A s^{-1}

B $mol\,dm^{-3}\,s^{-1}$

C $mol^{-1}\,dm^3\,s^{-1}$

D $mol^{-2}\,dm^6\,s^{-1}$

(1 mark)

The answer is C ✓

ⓔ The units of rate are $mol\,dm^{-3}\,s^{-1}$. This is the reason for answer B. The overall order of reaction (total of the orders) is 2. Dividing $mol\,dm^{-3}\,s^{-1}$ by $(mol\,dm^{-3})^2$ gives $mol^{(1-2)}\,dm^{(-3-(-6))}\,s^{-1} = mol^{-1}\,dm^3\,s^{-1}$. A would be the answer where the overall order of reaction is 1 and D where the overall order of reaction is 3.

Question 2

A graph of ln k against 1/T gave a gradient of −7142.9 and the graph intercepted the ln k axis at 20. The gas constant, R = 8.31 J K^{-1} mol^{-1}.

What is the activation energy for the reaction in kJ mol^{-1} to 3 significant figures?

A 0.166

B 59.4

C 59400

D 4.85×10^8

(1 mark)

The answer is B ✓

ⓔ The Arrhenius equation may be plotted graphically as $\ln k = E_a/RT + \ln A$. The vertical axis is ln k and the horizontal axis is 1/T. The gradient of the line is $-E_a/R$ and the intercept with the ln k axis is ln A. So the activation energy is determined from the gradient. $-E_a/R = -7142.9$. $E_a = 7142.9 \times R = 7142.9 \times 8.31 = 59\,357.5\,J\,mol^{-1}$. It is important to remember that this value is $J\,mol^{-1}$ and so dividing by 1000 gives 59.4 kJ mol^{-1}. Answer A would be obtained by confusing the intercept and the gradient and simply multiplying by R and dividing by 1000. Answer C has no dividing by 1000 step and answer D is e^{20} which is the method of calculating A from ln A from the intercept. Distractors will always be seemingly sensible processing of the numbers, so take care.

Question 3

A reacts with B to form C. A + 2B → C. The data in Table 4 are from some experiments into the rate of reaction.

Table 4

Experiment	Initial [A] ($mol\,dm^{-3}$)	Initial [B] ($mol\,dm^{-3}$)	Initial rate ($\times 10^{-4}\,mol\,dm^{-3}\,s^{-1}$)
1	0.127	0.240	1.99
2	0.127	0.480	7.99
3	0.254	0.240	4.00

(a) **Determine the order of reaction with respect to A and B.** (2 marks)

(b) **Write a rate equation for the reaction.** (1 mark)

(c) **Using experiment 3, calculate the value of the rate constant and state its units. Give your answer to 3 significant figures.** (3 marks)

(a) Order of reaction with respect to A = 1 ✓

Order of reaction with respect to B = 2 ✓

ⓔ As concentration of A doubles (experiment 1 to experiment 3) the rate doubles. So order of reaction with respect to A is 1. As the concentration of B doubles (experiment 1 to experiment 2) the rate quadruples (×4). So the order of reaction with respect to B is 2.

(b) rate = $k[A][B]^2$ ✓

ⓔ The rate equation simply gives the relationship between the concentrations of the reactants and the rate. Don't forget to include the rate constant, k. This should be a lower case k, not to be confused with an equilibrium constant, capital K.

(c) $4.00 \times 10^{-4} = k(0.254)(0.240)^2 k = \dfrac{4.00 \times 10^{-4}}{0.0146}$ ✓ = 0.0274 ✓ $mol^{-2}\,dm^6\,s^{-1}$ ✓

ⓔ Any experiment may be used to calculate a value for the rate constant, but if one is suggested, use it. Also make sure you take into account the orders when using the concentrations in the rate equation. The units are based on the overall order. Rate always has units of $mol\,dm^{-3}\,s^{-1}$, so dividing these units by $mol^3\,dm^{-9}$ gives $mol^{-2}\,dm^6\,s^{-1}$. 0.0273 is obtained if the entire calculation is carried out on your calculator. This would also be accepted. Show all steps clearly.

Question 4

The reaction between P and Q is shown by the equation:

$$2P + Q \rightarrow R + 3S$$

The rate equation is determined to be rate = $k[P]^2$

Table 5 gives data from a series of experiments into the rate of this reaction.

Table 5

Experiment	[P] ($mol\,dm^{-3}$)	[Q] ($mol\,dm^{-3}$)	Rate of reaction ($mol\,dm^{-3}\,s^{-1}$)
1	1.54×10^{-3}	2.10×10^{-3}	1.44×10^{-4}
2	2.31×10^{-3}	3.15×10^{-3}	to be calculated
3	to be calculated	5.25×10^{-3}	9.00×10^{-4}

(a) Calculate the rate of reaction in experiment 2. (1 mark)

From experiment 1 to experiment 2: [P] increases by a factor of 1.5 and [Q] increases by a factor of 1.5

Rate = $k[P]^2$, so change in [Q] has no effect.

As [P] increases ×1.5, rate increases $\times 1.5^2$ = rate ×2.25.

So new rate is $1.44 \times 10^{-4} \times 2.25 = 3.24 \times 10^{-4}$ ✓

ⓔ It is important to be able to manipulate these powers. When a concentration is multiplied by a factor — call it factor *p* in this case — for the ×1.5 increase in [P], the rate increases by that factor raised to the order of the reactant. As P has order 2, $(1.5)^2 = 2.25$, so the rate increases by a factor of 2.25. It is important to note that as Q is zero order, any change in Q has no effect on the rate of reaction.

(b) Calculate the concentration of P in experiment 3. (1 mark)

From experiment 1 to experiment 3 the rate increases by a factor of 6.25 ($9.00 \times 10^{-4}/1.44 \times 10^{-4}$).

[P] increases by the factor which is the square root of this, which is $\sqrt{6.25} = 2.5$.

So [P] from experiment 1 to 3 should be ×2.5.

$1.54 \times 10^{-3} \times 2.5 = 3.85 \times 10^{-3}$ ✓

ⓔ Again [Q] has no effect on the rate of this reaction.

(c) Use experiment 1 to calculate a value for the rate constant and state its units. Give your answer to 3 significant figures. (3 marks)

$1.44 \times 10^{-4} = k(1.54 \times 10^{-3})^2$

$k = \frac{1.44 \times 10^{-4}}{(1.54 \times 10^{-3})^2}$ ✓ = 60.7 ✓ $mol^{-1}\,dm^3\,s^{-1}$ ✓

(d) Use the Arrhenius equation to calculate a value for the Arrhenius constant A given that the activation energy for the reaction is +84.6 kJ mol^{-1} at 300 K and the gas constant R = 8.31 J K^{-1} mol^{-1}. Give your answer to 3 significant figures. (2 marks)

$$k = Ae^{-\frac{E_a}{RT}}$$

$$60.7 = Ae^{-\frac{84600}{8.31\times300}} ✓$$

$$60.7 = A(1.83 \times 10^{-15})$$

$$A = \frac{60.7}{1.83 \times 10^{-15}} = 3.32 \times 10^{16} ✓$$

ⓔ Again remember to convert E_a to J mol^{-1} for use in the expression. The clue is that R is given in J K^{-1} mol^{-1}, so this leaves the power unitless if E_a is in J mol^{-1}. Use A to check the calculation back to ensure you obtain a value of around 60.7 for k.

Equilibrium constant K_p for homogeneous systems

Question 1

A gaseous mixture at equilibrium contains 0.250 moles of hydrogen and 0.175 moles of nitrogen and 0.345 moles of ammonia. The total pressure is 110 kPa.

$$N_2(g) + 3H_2(g) \rightleftharpoons 2NH_3(g)$$

What is the partial pressure of ammonia in this mixture?

A 25.0 kPa

B 35.7 kPa

C 49.3 kPa

D 89.6 kPa

(1 mark)

The answer is C ✓

ⓔ Mole fractions are determined by dividing the moles at equilibrium by the total equilibrium moles (0.250 + 0.175 + 0.345 = 0.770). The partial pressure is calculated by multiplying the mole fraction by the total pressure (110 kPa). A is the partial pressure of nitrogen, B is the partial pressure of hydrogen and D is the answer which would be obtained if the 2 in the equation for ammonia were to be used incorrectly. When amounts, in moles, at equilibrium are given, there is no need to consider the balancing numbers in the equation.

Question 2

For the reaction:

$$N_2O_4(g) \rightleftharpoons 2NO_2(g)$$

1.00 moles of N_2O_4 are allowed to reach equilibrium at 500 K. At equilibrium 0.240 moles of N_2O_4 remain. The total pressure is 200 kPa.

(a) Calculate the mole fractions of N_2O_4 and NO_2 at equilibrium. Give your answer to 3 significant figures. (4 marks)

Moles of N_2O_4 at equilibrium = 0.240 mol

Moles of NO_2 at equilibrium = 1.52 mol ✓

Total equilibrium moles = 0.24 + 1.52 = 1.76 mol ✓

Mole fraction of N_2O_4 = 0.240/1.76 = 0.136 ✓

Mole fraction of NO_2 = 1.52/1.76 = 0.864 ✓

ⓔ The moles at equilibrium are calculated using the initial 0.240 moles of N_2O_4. The amount, in moles, of NO_2 present at equilibrium is $2 \times 0.760 = 1.52$ mol based on the balancing numbers in the equation for the reaction. The total equilibrium moles is 0.240 + 1.52 = 1.76 moles. Mole fractions are calculated by dividing the moles by the total equilibrium moles.

(b) Calculate the partial pressures of N_2O_4 and NO_2 at equilibrium. Give your answer to 3 significant figures. (2 marks)

Partial pressure of N_2O_4 = 0.136×200 = 27.2 kPa ✓

Partial pressure of NO_2 = 0.864×200 = 173 kPa ✓

ⓔ The partial pressures are calculated by multiplying the moles fractions by the total pressure.

(c) Calculate a value for K_p at 500 K. State its units. Give your answer to 3 significant figures. (3 marks)

$$K_p = \frac{(pNO_2)^2}{(pN_2O_4)} \checkmark = \frac{(173)^2}{(27.2)} = 1100 \checkmark \text{ kPa} \checkmark$$

ⓔ The calculation of the K_p is from partial pressures. If K_p has no units, then equilibrium moles or mole fractions may be used to calculate K_p.

Question 3

Hydrogen iodide decomposes into hydrogen and iodine according to the equilibrium:

$2HI(g)\ \ H_2(g) + I_2(g)$

0.0248 moles of hydrogen iodide were placed in a sealed container at 350 K. Calculate a value for K_p at 350 K if 0.0220 moles of hydrogen iodide are present in the equilibrium mixture. Give your answer to 3 significant figures. (4 marks)

Moles of HI which react = 2.80×10^{-3} mol ✓

Moles of H_2 and I_2 formed = 1.40×10^{-3} mol ✓

$$K_p = \frac{(pH_2)(pI_2)}{(pHI)^2} = \frac{(1.40 \times 10^{-3})^2}{(0.0220)^2} ✓ = 4.05 \times 10^{-3} ✓$$

or, using mole fractions:

Total equilibrium moles = $0.0220 + 1.40 \times 10^{-3} + 1.40 \times 10^{-3} = 0.0248$ mol

Mole fraction of HI = $\frac{0.0220}{0.0248} = 0.887$ ✓

Mole fraction of H_2 = mole fraction of I_2 = $\frac{1.4 \times 10^{-3}}{0.0248} = 0.0565$ ✓

$$K_p = \frac{(pH_2)(pI_2)}{(pHI)^2} = \frac{(0.0565)^2}{(0.887)^2} ✓ = 4.06 \times 10^{-3} ✓$$

ⓔ As K_p has no units, either equilibrium moles or mole fractions may be used to calculate the value of K_p. Note that if the question requires you to calculate mole fractions, you should do this. Also some texts will show *P* as the total pressure multiplied by the mole fractions to calculate the partial pressures, but the *P* will cancel out in the K_p expression. The slight difference in the answers is due to rounding in the mole fractions.

Question 4

Sulfur dioxide reacts with oxygen to form sulfur trioxide according to the equation:

$2SO_2(g) + O_2(g) \rightleftharpoons 2SO_3(g)$

0.115 moles of sulfur dioxide were mixed with 0.210 moles of oxygen in a container at 700 K. At equilibrium 68% of the sulfur dioxide had reacted. K_p for the reaction at 700 K is 0.0651 kPa^{-1}. Calculate the total pressure in this reaction and state its units. Give your answer to 3 significant figures. (5 marks)

$$2SO_2 \quad + \quad O_2 \quad \rightleftharpoons \quad 2SO_3$$

	$2SO_2$	O_2	$2SO_3$
Initial moles	0.115	0.210	0
Reacting moles	−0.0782 (68% of 0.115)	−0.0391	+0.0782
Equilibrium moles	0.0368	0.171	0.0782
Mole fraction	$\frac{0.0368}{0.286} = 0.129$	$\frac{0.171}{0.286} = 0.598$	$\frac{0.0782}{0.286} = 0.273$
Partial pressure	$0.129P$ ✓	$0.598P$ ✓	$0.273P$ ✓

Total equilibrium moles = 0.286

$$K_p = \frac{(pSO_3)^2}{(pSO_2)^2(pO_2)} = \frac{(0.273P)^2}{(0.129P)^2(0.598P)} \checkmark$$

$$= \frac{0.0745P^2}{9.95 \times 10^{-3}P^3}$$

$$\text{So } \frac{0.0745}{9.95 \times 10^{-3}P} = 0.0651\,\text{kPa}^{-1}$$

$$P = \frac{0.0745}{9.95 \times 10^{-3} \times 0.0651} = 115\,\text{kPa} \checkmark$$

ⓔ This is a complex calculation which takes time. If you have time, you can put the 115 kPa answer back into the calculation to make sure you get the same value for K_p.

Electrode potentials and electrochemical cells

Question 1

Using the standard electrode potentials below, choose the oxidising agent capable of oxidising vanadium from the +2 to the +4 oxidation state but not to the +5 oxidation state.

	$E^\ominus$/V
$Cl_2(aq) + 2e^- \rightleftharpoons Cl^-(aq)$	+1.36
$I_2(aq) + 2e^- \rightleftharpoons 2I^-(aq)$	+0.54
$SO_4^{2-}(aq) + 2H^+(aq) + 2e^- \rightleftharpoons 2H_2O(l) + SO_2(g)$	+0.17
$VO^{2+}(aq) + 2H^+(aq) + e^- \rightleftharpoons V^{3+}(aq) + H_2O(l)$	+0.32
$VO_2^+(aq) + 2H^+(aq) + e^- \rightleftharpoons VO^{2+}(aq) + H_2O(l)$	+1.00
$V^{3+}(aq) + e^- \rightleftharpoons V^{2+}(aq)$	−0.26
$Sn^{2+}(aq) + 2e^- \rightleftharpoons Sn(s)$	−0.15

A chlorine　　**C** iodine

B Sn^{2+} ions　　**D** sulfate ions　　(1 mark)

The answer is C ✓

ⓔ It is important to be able to recognise vanadium in its various oxidation states: VO_2^+ represents vanadium in the +5 oxidation state; VO^{2+} represent vanadium in its +4 oxidation state; V^{3+} and V^{2+} speak for themselves in the +3 and +2 oxidation states respectively. The electrode potentials are all stated as reduction reactions. The strongest oxidising agent will be found on the left-hand side of these half-equations. Iodine reduces to iodide with a potential of +0.54 V.

	Oxidation state change of V	Oxidation potential (V)	$Cl_2 \rightarrow 2Cl^-$ (+1.36)	$Sn^{2+} \rightarrow Sn$ (−0.15)	$I_2 \rightarrow 2I^-$ (+0.54)	$SO_4^{2-} \rightarrow SO_2$ (+0.17)
1	+2 → +3	+0.26	✓	✓	✓	✓
2	+3 → +4	−0.32	✓	✗	✓	✗
3	+4 → +5	−1.00	✓	✗	✗	✗

A ✓ in the table shows a positive emf, a ✗ shows a negative emf. The only reduction reaction that will make the emf for the first and second reactions positive is $I_2 \rightarrow 2I^-$ (+0.54 V). Chlorine would oxidise vanadium from +2 to +5. Sn^{2+} and sulfate ions would oxidise vanadium from +2 to +3 but no further.

Question 2

Figure 2 shows an electrochemical cell.

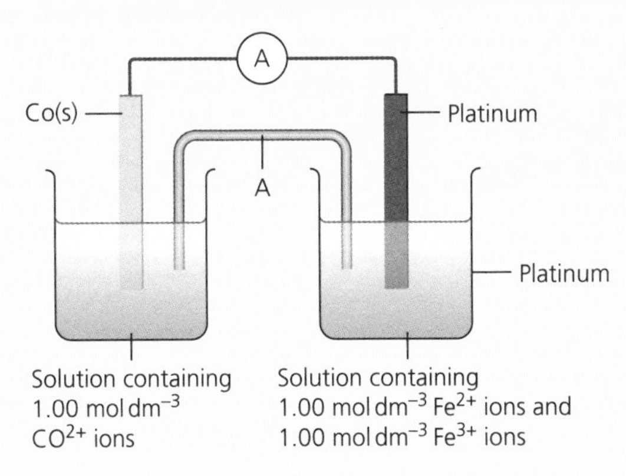

Figure 2

The half-equations and electrode potentials are:

$Co^{2+}(aq) + 2e^- \rightarrow Co(s)$ −0.28 V

$Fe^{3+}(aq) + e^- \rightarrow Fe^{2+}(aq)$ to be determined

(a) Write the conventional cell representation for this cell. (2 marks)

Co | Co^{2+} || Fe^{3+}, Fe^{2+} | Pt ✓✓

e The conventional cell representation is sometimes called cell notation and has the oxidation on the left and the reduction on the right. The || in the middle represents the salt bridge. Any substances in different phases (states) are separated by a phase boundary line (|). Substance such as Fe^{3+} and Fe^{2+} are in the same state and so are separated by a comma. Pt is included as the electrode in the iron half cell at the right, again separated by a phase boundary line as it is a solid and the other ions are in solution.

(b) Write an overall equation for the reaction occurring. (2 marks)

$Co^{2+} + 2Fe^{3+} \rightarrow Co + 2Fe^{2+}$ ✓✓

e Combining the oxidation and reduction half-equations requires the balancing of the electrons. So first reverse the cobalt half-equation to make it the oxidation and multiply the reduction iron half-equation by 2. Then simply add them together and cancel out the $2e^-$ on each side.

(c) The emf of the cell is 1.05 V. Calculate the standard electrode potential for the half-equation $Fe^{3+}(aq) + e^- \rightarrow Fe^{2+}(aq)$ (2 marks)

$1.05 = x - (-0.28)$ ✓

$x = 1.05 - 0.28 = +0.77\,V$ ✓

e The skill here is remembering that the emf is determined by adding together the reduction and oxidation potentials. One of the potentials has its sign changed. In this example the cobalt is oxidised, so the iron potential we are calculating is the reduction one. This can also be achieved by using emf = $E_{rhs} - E_{lhs}$ as the $- E_{lhs}$ changes the sign of the oxidation reaction for you. This is the method used in the answer.

(d) What direction do the electrons flow in the cell? (1 mark)

From left to right/cobalt electrode to iron electrode ✓

e Electrons are released when the cobalt is oxidised. These electrons flow around the circuit to the iron half cell and the Fe^{3+} is reduced to Fe^{2+}. So the electrons always flow from the oxidation half cell to the reduction half cell.

(e) What change, if any, would be observed in the mass of the cobalt? (1 mark)

Decrease ✓

e As cobalt is oxidised the reaction which occurs in the left-hand cell is $Co \rightarrow Co^{2+} + 2e^-$. The cobalt will go into solution and so the mass of the cobalt will decrease.

(f) Identify the negative electrode. (1 mark)

Cobalt electrode ✓

ⓔ NEGATOX is used to identify the negative electrode. The negative electrode is where an oxidation reaction occurs. The oxidation occurs at the cobalt electrode.

Question 3

Use the standard electrode potentials given in Table 6 for this question.

Table 6

Half reaction	*E*/V
$F_2(g) + 2e^- \rightarrow 2F^-(aq)$	+2.87
$Cl_2(g) + 2e^- \rightarrow 2Cl(aq)$	+1.36
$Br_2(l) + 2e^- \rightarrow 2Br^-(aq)$	+1.07
$Fe^{3+}(aq) + e^- \rightarrow Fe^{2+}(aq)$	+0.77
$I_2(s) + 2e^- \rightarrow 2I^-(aq)$	+0.54
$Cu^{2+}(aq) + 2e^- \rightarrow Cu(s)$	+0.34
$Sn^{4+}(aq) + 2e^- \rightarrow Sn^{2+}(aq)$	+0.14
$2H^+(aq) + 2e^- \rightarrow H_2(g)$	0.00
$Pb^{2+}(aq) + 2e^- \rightarrow Pb(s)$	−0.13
$Fe^{2+}(aq) + 2e^- \rightarrow Fe(s)$	−0.44
$Zn^{2+}(aq) + 2e^- \rightarrow Zn(s)$	−0.76

(a) Explain why a reaction occurs between chlorine and iron(II) ions in solution but not between iodine and iron(II) ions. (2 marks)

$E^\ominus$ $(Cl_2/Cl^-) > E^\ominus$ (Fe^{3+}/Fe^{2+}) or emf is +0.59 V ✓

$E^\ominus$ $(I_2/I^-) < E^\ominus$ (Fe^{3+}/Fe^{2+}) or emf is −0.23 V ✓

ⓔ Calculation of the emf for the reaction between Cl_2 and Fe^{2+} ions gives a positive value or you can state that the reduction of Cl_2 has a greater electrode potential than the reduction of Fe^{3+} to Fe^{2+}, so Cl_2 will oxidise Fe^{2+} to Fe^{3+}. However, a negative emf is calculated for the reaction of I_2 with Fe^{2+}, or the potential for the reduction of Fe^{3+} to Fe^{2+} is greater than the potential for the reduction of I_2 to I^-.

(b) Identify the strongest oxidising agent from Table 6. (1 mark)

F_2/fluorine ✓

ⓔ Fluorine is the species in the table that is most easily reduced, because it has the most positive value for its reduction to fluoride ions. Zinc is the strongest reducing agent in the table, because it is most easily oxidised since it has the most positive value for its oxidation if the half-equations were reversed and the signs changed.

(c) A conventional cell representation is Pb | Pb^{2+} || Cu^{2+} | Cu.

(i) Calculate the emf of the cell. (1 mark)

0.47 V ✓

ⓔ Again this type of calculation can be done by simply reversing one of the equations (in this case the lead one) and changing the sign of the electrode potential and adding them together. Alternatively the equation below can be used:

$$\text{emf} = E_{rhs} - E_{lhs}$$

$$\text{emf} = +0.34 - (-0.13) = +0.47\,V$$

(ii) Identify the positive electrode. (1 mark)

Copper electrode ✓

ⓔ Using NEGATOX the negative electrode is the one where oxidation occurs, so the positive electrode is the one where reduction occurs. The reduction electrode is the copper electrode.

Question 4

The half-equations for the lithium ion cell are given below:

$Li^+ + e^- \rightarrow Li$ $\quad E^\ominus = -3.03\,V$

$Li^+ + CoO_2 + e^- \rightarrow Li^+[CoO_2]^-$ $\quad E^\ominus = +0.67\,V$

(a) What is the change in oxidation state of the cobalt in the second half-equation? (2 marks)

+4 ✓ to +3 ✓

ⓔ The oxidation state of oxygen is −2, so Co in CoO_2 is +4, whereas in CoO_2^- cobalt has an oxidation state of +3.

(b) Write an overall equation for the reaction that occurs. (2 mark)

$Li + CoO_2 \rightarrow Li^+[CoO_2]^-$ ✓✓

ⓔ Again combining the equations by reversing the top one and adding them together. The Li^+ and e^- can be cancelled out.

(c) Calculate the voltage that can be supplied by the cell. (1 mark)

3.70 V ✓

ⓔ As the top equation becomes the oxidation one, the overall emf of the cell, which is the same as the voltage supplied, is 3.03 + 0.67 = 3.70 V.

(d) Suggest why water is not used as the solvent in this type of cell. (1 mark)

Lithium reacts with water ✓

ⓔ This type of 'suggest' question tests your ability to think outside the topic and apply knowledge.

Acids and bases

Question 1

Which one of the pH curves in Figure 3 would show the change in pH when 0.100 mol dm^{-3} sodium hydroxide solution is added to 25.0 cm^3 of 0.100 mol dm^{-3} hydrochloric acid? (1 mark)

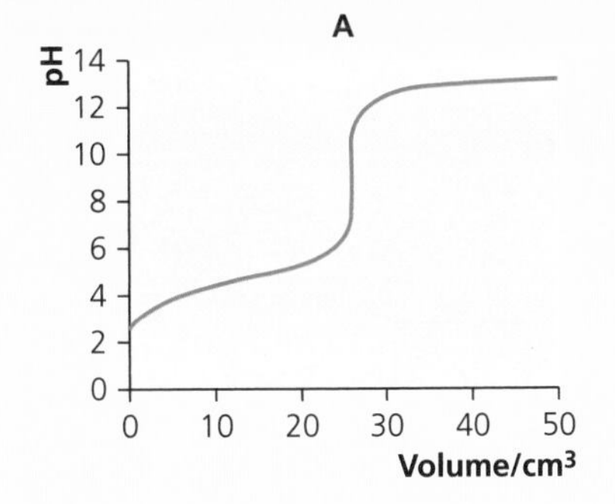

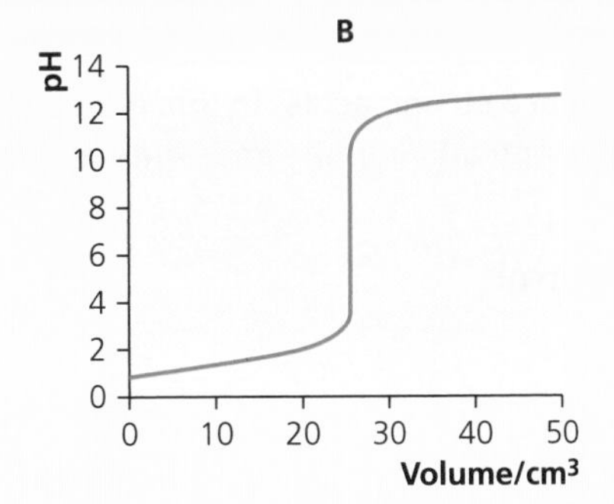

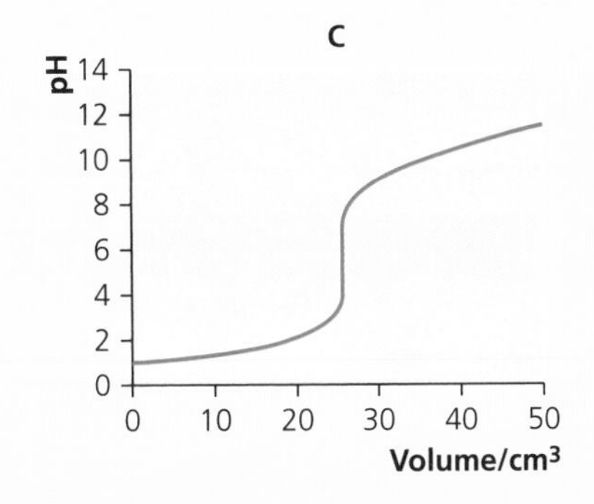

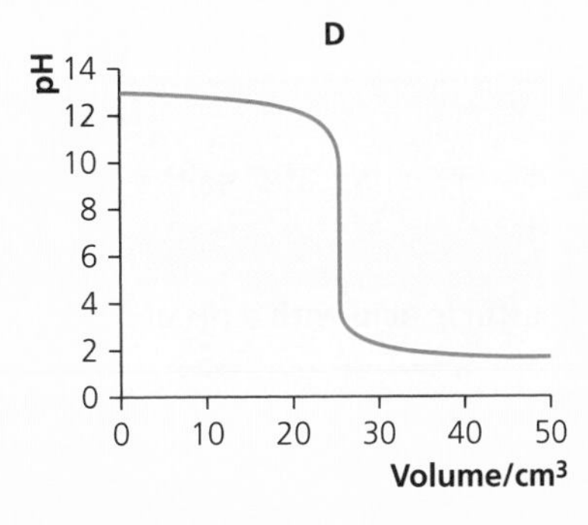

Figure 3

The answer is B ✓

e The pH curve must go from low to high pH. As it is a strong acid–strong base titration, the vertical region should go from around pH 3 to 10. The volume at neutralisation will be $25.0\,cm^3$. Curve A would be for the addition of a strong base to a weak acid; curve C would be for the addition of a weak base to a strong acid; curve D would be for the addition of a weak acid to a strong base.

Question 2

What is the pH of a solution of ethanoic acid ($K_a = 1.74 \times 10^{-5}\,mol\,dm^{-3}$) of concentration $0.224\,mol\,dm^{-3}$? (1 mark)

A 0.65 **C** 4.76

B 2.71 **D** 5.41

The answer is B ✓

e The $[H^+]$ of a weak acid = $\sqrt{K_a \times [\text{weak acid}]}$. The pH is determined using pH $= -\log_{10}[H^+]$. $[H^+]$ for this acid = $\sqrt{1.74 \times 10^{-5} \times 0.224} = 1.97 \times 10^{-3}\,mol\,dm^{-3}$. pH $= -\log_{10}(1.97 \times 10^{-3}) = 2.71$. Answer A is obtained if the concentration of $[H^+]$ is taken as $0.224\,mol\,dm^{-3}$; answer C is obtained if $-\log_{10}K_a$ is calculated; answer D is obtained if the square root is not taken.

Question 3

Hydrochloric acid and sulfuric acid are strong acids. In this question give all values of pH to 2 decimal places and all volumes and concentrations to 3 significant figures.

(a) Write an expression for the term pH. (1 mark)

pH $= -\log_{10}[H^+]$ ✓

(b) Calculate the pH of $0.125\,mol\,dm^{-3}$ hydrochloric acid. (2 marks)

$[H^+] = 0.125\,mol\,dm^{-3}$ ✓ pH $= -\log_{10}[H^+] = -\log_{10}(0.125) = 0.90$ ✓

e Hydrochloric acid is a strong acid and monoprotic, so the concentration of H^+ is the same as the concentration of HCl.

(c) Calculate the concentration of sulfuric acid with a pH of 0.50. (2 marks)

$[H^+] = 10^{-0.50} = 0.316\,mol\,dm^{-3}$ ✓ $[H_2SO_4] = 0.158\,mol\,dm^{-3}$ ✓

e The concentration of $[H^+]$ is calculated using the pH as $[H^+] = 10^{-pH}$. As H_2SO_4 is a strong diprotic acid the concentration of the acid is half the concentration of hydrogen ions.

(d) Calculate the volume of water that would be added to 25.0 cm^3 of 0.125 mol dm^{-3} hydrochloric acid to give a solution with pH 1.54. (4 marks)

moles of H^+ in original solution = $\frac{(25.0 \times 0.125)}{1000}$ = 3.125×10^{-3} mol ✓

$[H^+]$ in new solution = $10^{-1.54}$ = 0.0288 mol dm^{-3} ✓

$\frac{3.125 \times 10^{-3}}{V} \times 1000 = 0.0288$, where V is new total volume of solution

$V = \frac{3.125 \times 10^{-3}}{0.0288} \times 1000 = 108.5$ cm^3 ✓

volume of water added = 108.5 − 25 = 83.5 cm^3 ✓

e Answers in the range 83.0 cm^3 to 83.7 cm^3 would be acceptable due to rounding to 3 significant figures during the calculation. It is acceptable to work to 4 significant figures during a calculation as long as the answer is given to the required number of significant figures. The dilution of the number of moles of H^+ to a new concentration can be determined by dividing the moles of H^+ in the original solution by the volume of the solution and multiplying by 1000.

(e) 30.0 cm^3 of 0.170 mol dm^{-3} sodium hydroxide solution is added to 25.0 cm^3 of 0.125 mol dm^{-3} sulfuric acid. Calculate the pH of the solution formed. (6 marks)

Moles of NaOH = $\frac{30.0 \times 0.170}{1000}$ = 5.10×10^{-3} mol ✓

Moles of H_2SO_4 = $\frac{25.0 \times 0.125}{1000}$ = 3.125×10^{-3} mol ✓

Ratio of NaOH:H_2SO_4 = 2:1

Moles of H_2SO_4 in excess = 5.75×10^{-4} mol ✓

Total volume = 55.0 cm^3

New $[H_2SO_4]$ = $\frac{5.75 \times 10^{-4}}{55.0} \times 1000$ = 0.0105 mol dm^{-3} ✓

$[H^+] = 2 \times [H_2SO_4]$ = 0.0210 mol dm^{-3} ✓

pH = $-\log_{10}[H^+] = -\log_{10}(0.0210) = 1.68$ ✓

e This is a complex calculation due to the 2:1 ratio of NaOH to H_2SO_4. Dividing 5.10×10^{-3} by 2 gives 2.55×10^{-3} and this is subtracted from the moles of H_2SO_4 (3.125×10^{-3}) to determine the moles of H_2SO_4 in excess. The total volume is 55.0 cm^3 as this is the total volume of the two solutions when mixed. This allows the concentration of H_2SO_4 to be calculated. As H_2SO_4 is a strong diprotic acid, $[H^+] = 2 \times [H_2SO_4]$. The pH is calculated in the normal way using pH = $-\log_{10}[H^+]$. Watch out for calculations where the base is in excess, as K_w is required to calculate $[H^+]$.

Question 4

(a) **20.0 cm^3 of 0.0500 $mol\,dm^{-3}$ sodium hydroxide solution were added to 30.0 cm^3 of 0.0500 $mol\,dm^{-3}$ ethanoic acid ($K_a = 1.74 \times 10^{-5}\,mol\,dm^{-3}$). Calculate the pH of the buffer solution formed. Give your answer to 2 decimal places.** (6 marks)

Moles of $CH_3COOH = \dfrac{30.0 \times 0.0500}{1000} = 1.50 \times 10^{-3}\,mol$

Moles of $NaOH = \dfrac{20.0 \times 0.0500}{1000} = 1.00 \times 10^{-3}\,mol$

$CH_3COOH + NaOH \rightarrow CH_3COONa + H_2O$

Moles of ethanoic acid remaining $= 1.50 \times 10^{-3} - 1.00 \times 10^{-3} = 5.00 \times 10^{-4}\,mol$ ✓

Moles of sodium ethanoate formed $= 1.00 \times 10^{-3}\,mol$ ✓

Total volume of solution $= 50.0\,cm^3$

New $[CH_3COOH] = \dfrac{5.00 \times 10^{-4}}{50.0} \times 1000 = 0.0100\,mol\,dm^{-3}$ ✓

New $[CH_3COO^-] = \dfrac{1.00 \times 10^{-3}}{50.0} \times 1000 = 0.0200\,mol\,dm^{-3}$ ✓

$K_a = \dfrac{[H^+][CH_3COO^-]}{[CH_3COOH]}$ $[H^+] = \dfrac{K_a[CH_3COOH]}{[CH_3COO^-]}$

$[H^+] = \dfrac{1.74 \times 10^{-5} \times 0.0100}{0.0200} = 8.70 \times 10^{-6}\,mol\,dm^{-3}$ ✓

$pH = -\log_{10}[H^+] = -\log_{10}(8.70 \times 10^{-6}) = 5.06$ ✓

e Buffer calculations are never easy, but it is important that you aim to discover the new concentration of the weak acid and its anion in the solution, taking into account the new total volume of the solution.

(b) **5.0 cm^3 of a 0.0105 $mol\,dm^{-3}$ solution of hydrochloric acid are added to the buffer solution in (a). Calculate the pH of the buffer solution after the addition of the acid.** (6 marks)

Moles of H^+ added $= \dfrac{5.00 \times 0.0105}{1000} = 5.25 \times 10^{-5}\,mol$ ✓

New moles of $CH_3COOH = 5.00 \times 10^{-4} + 5.25 \times 10^{-5} = 5.525 \times 10^{-4}\,mol$ ✓

New moles of $CH_3COO^- = 1.00 \times 10^{-3} - 5.25 \times 10^{-5} = 9.475 \times 10^{-4}\,mol$ ✓

New total volume $= 50 + 5 = 55\,cm^3$

New $[CH_3COOH] = \dfrac{5.525 \times 10^{-4}}{55.0} \times 1000 = 0.0100\,mol\,dm^{-3}$ ✓

New $[CH_3COO^-] = \dfrac{9.475 \times 10^{-4}}{55.0} \times 1000 = 0.0172\,mol\,dm^{-3}$ ✓

$$K_a = \frac{[H^+][CH_3COO^-]}{[CH_3COOH]} \quad [H^+] = \frac{K_a[CH_3COOH]}{[CH_3COO^-]}$$

$$[H^+] = \frac{1.74 \times 10^{-5} \times 0.0100}{0.0172} = 1.01 \times 10^{-5}\,mol\,dm^{-3}\ ✓$$

$$pH = -\log_{10}[H^+] = -\log_{10}(3.68 \times 10^{-5}) = 4.99\ ✓$$

e A buffer maintains an almost constant pH on the addition of small quantities of an acid or a base. The pH of the buffer should not be very different on the addition of the acid or the base. Remember that added H^+ increases the HA and decreases the A^-, but the addition of a base decreases the HA and increases the A^-. The total volume should be taken into account, though using the moles of HA and A^- will give the same answer.

Knowledge check answers

1 The standard enthalpy change of atomisation is the enthalpy change when 1 mole of gaseous atoms is formed from the element in its standard state under standard conditions.
2 $CaO(s) \rightarrow Ca^{2+}(g) + O^{2-}(g)$
3 $-15\,kJ\,mol^{-1}$
4 Standard change in entropy and $J\,K^{-1}\,mol^{-1}$
5 $kJ\,mol^{-1}$
6 350.9 K
7 $mol^{-2}\,dm^6\,s^{-1}$
8 $J\,mol^{-1}$
9 $K_p = \frac{(pNO_2)^2}{(pN_2O_4)}$
10 1
11 Position of equilibrium moves to right
No change for K_p
12 100 kPa, 298 K, any solutions $1.00\,mol\,dm^{-3}$
13 Zinc
14 +3
15 Ni +3 to +2
16 HCO_3^-
17 $pH = -\log_{10}[H^+]$
18 −0.71
19 $K_a = \frac{[HCOO^-][H^+]}{[HCOOH]}$
20 2.84
21 2.24
22 Pink
23 Maintains an almost constant pH on addition of small amounts of acid or base

Index

Note: page numbers in **bold** indicate key term definitions.

A

acid dissociation constant (K_a) 59
acids and bases 52–71
  Brønsted–Lowry theory 52–53
  buffers 66–71
  classification of 53–54
  dilutions 60–61
  ionic product of water 55–56
  neutralisations 61–63
  pH of strong acids 54–55
  pH of strong alkalis 56–57
  pH of weak acids 57–60
  pH calculation 54
  pH curves 63–66
  questions and answers 89–93
activation energy (E_a)
  Arrhenius equation 29
  and rate of reaction 24
Arrhenius equation 29–30

B

batteries (cells) 49–52
bond dissociation enthalpy **7**
Born–Haber cycles 6–11, 13
Brønsted–Lowry acid **52**
Brønsted–Lowry base **52**
buffers 66–71
  acidic 67
  action of 66–67
  addition of acid or alkali to 70–71
  basic 67
  calculating pH of 68–70
  preparation of acidic 67

C

changes of state, entropy of 14–15
concentration against time graphs, reaction rates 28–29
continuous rate monitoring, rate of reaction 27–29

D

dilutions, strong acids and bases 60–61

E

electrochemical cells
  basic electrochemistry of 49
  commercial application 49
  fuel cells 51–52
  positive emf and feasibility of reaction 45–46
  primary cells 49–50
  risks and benefits 52
  secondary cells 50–51
electrode potential **40**
electrode potentials 39–48
  applied to vanadium chemistry 47–48
  feasibility of reactions 45–46
  oxidising agents and reducing agents 46
  questions and answers 84–89
  reactivity order 46
  standard electrode potentials 40–45
electromotive force (emf) **40**
  calculation 44
  positive emf and feasibility of reaction 40, 45–46
enthalpy of solution **12**
  calculating 13
  hydration enthalpy 12–13
entropy change ($\Delta S^{\ominus}$) 14–16
  calculations 16
  changes of state 14–15
entropy (S) **15**
  values for various substances 15–16
equilibrium constant (K_p) for homogeneous systems 32–39
  calculating and using 33–37
  expressions and units of 32
  mole fractions and partial pressures 32–33
  questions and answers 81–84
  temperature affecting value of 39
  total pressure calculation 38
experimental value for lattice enthalpy 11–12

F

feasibility of reactions 45–46
  factors affecting 18
  Gibbs free energy, $\Delta G^{\ominus}$ 17
  and positive emf 40, 45–46
  and temperature 18–20
first electron affinity **8**
first ionisation energy **7**
fuel cells 51–52

G

gas syringes 26
gas volume–time graphs 26–27
Gibbs free energy change ($\Delta G^{\ominus}$) 14, 17
  and feasibility of a reaction 18
  graphs of $\Delta G^{\ominus}$ against temperature 19–20
  physical changes related to 18
group 1 halides, Born–Haber cycle 8–9
group 2 halides, Born–Haber cycle 9–10

H

half cells 40
  combination of 41–42
  conventional representation 43–44
  left-hand cell or right-hand cell 42–43
  non-standard conditions 43
  phase boundaries 44–45
Henderson–Hasselbalch equation 70
Hess's law 6, 7, 9
hydration enthalpy **12–13**
hydrogen fuel cells 51, 52

I

indicators 66
initial rate monitoring, rate of reaction 26–27
ionic product of water 55–56

K

K_a (acid dissociation constant) 57–58, 68

K_p *see* equilibrium constant for homogeneous systems
K_w (ionic product of water) 55–57, 61

L
lattice enthalpies
  patterns 11
  theoretical and experimental values 11–12
lead–acid cell 50–51
lithium ion cells 51

M
methyl orange 66
methyl red 66
mole fractions 32–33
  calculations 35, 36–37

N
NEGATOX mnemonic 49
neutralisations 61–63
nickel cadmium cell 50

O
order of reaction **21**
  determining and using 21–24
  linked to mechanism 30–31
overall order of reaction **21**
oxides, Born–Haber cycle 10–11
oxidising agents 46

P
partial pressures 32–33
  calculations 33–34, 37
perfect ionic model 11–12
pH of strong acids 54–55
pH of strong alkalis 56–57
pH of water 56
pH of weak acids 57–60
pH calculation 54
pH curves 63–66
phase boundaries 44
phenolphthalein 66
pK_a 58–60
position of equilibrium and K_c 39
potassium chloride (KCl), Born–Haber cycle 6–7
primary cells 49–50

R
rate of reaction, methods of determining 25–29
  continuous rate monitoring 27–29
  initial rate monitoring 26–27
  rate measurements 25–26
rate constant (k) 20–21
  Arrhenius equation 29–30
  factors affecting 24–25
  units of 21
rate determining step 30–31
rate equations 20–31
  questions and answers 78–81
reactivity–feasibility link 46
reducing agents 46
required practicals
  investigating pH changes 66
  measuring emf of electrochemical cell 49
  measuring rate of reaction 31

S
secondary cells 50–51
standard electrode potential **40**
standard electrode potentials 40–45
  combination of half cells 41–42
  conventional cell representation 43–44
  left-hand cell or right-hand cell 42–43
  non-standard conditions in a cell 43
  phase boundaries 44–45
  standard hydrogen electrode 41
standard enthalpy change of atomisation **7**
standard enthalpy change of formation **6**
standard entropy ($S^{\ominus}$) 15
standard hydrogen electrode 41
standard lattice enthalpy **6**
strong acids 53
  dilutions of 60–61
  pH of 54–55
strong alkalis, pH of 56–57

T
temperature
  effect on value of K_p 39
  graphs of $\Delta G^{\ominus}$ against 19–20
  and rate constant 24–25
theoretical value for lattice enthalpy 11–12
thermodynamics 6–20
  Born–Haber cycles 6–14
  Gibbs free energy and entropy 14–20
  questions and answers 73–77
titration 63–64
  curves, sketching 64–65
  indicators for 66
total pressure calculation 38

V
vanadium chemistry, electrode potentials applied to 47–48

W
water, calculating pH of 56
weak acids 53–54
  and buffers 67–69
  pH and pK_a 57–60